H. G Stevens

Stevens Mechanical Catechism

H. G Stevens
Stevens Mechanical Catechism
ISBN/EAN: 9783743687240

Printed in Europe, USA, Canada, Australia, Japan

Cover: Foto ©berggeist007 / pixelio.de

More available books at **www.hansebooks.com**

STEVENS

Mechanical Catechism

FOR

STATIONARY AND MARINE ENGINEERS, FIREMEN
ELECTRICIANS, MOTOR-MEN, ICE-MACHINE
MEN AND MECHANICS IN GENERAL

PRACTICAL KNOWLEDGE

IN EVERY BRANCH OF MECHANICAL INDUSTRY

Full Information on Water, Steam, Fire, Smoke, Electricity, Horse-Power, Refrigeration, Liquid Air. Exact Description of, and Directions for the Care of Boilers, Grates, Engines, Slide Valve, Safety Valves, Injectors, Pumps, Steam Gauges, Lubricators, Eccentric, Link Motion, Indicator, Ammonia Compressor, Brine and Direct Expansion Systems, Lathe, Tools, Dynamo, Batteries, Parallel and Series Wiring, Three-Wire System, Motors, Controller, Electric Heating, House Wiring, Traction Engine. Thorough Instruction in Calculation of Horse-Power, Pulley-Speed, Lathe-Gearing, Square Root, Leverage, Tensile Strength, etc. Introduction to Algebra. Systematic descriptions alternate with elaborate sets of Questions and Answers in plainest English. Numerous tables and original diagrams make the book interesting as well as instructive. Valuable Recipes and Hints for all sorts of Emergencies, many of them especially selected for this work.

OVER 240 SECTIONAL CUTS AND ILLUSTRATIONS

BY

H. G. STEVENS, M.E.E.

Copyright 1899, by WM. H. LEE

CHICAGO
LAIRD & LEE

TABLE OF CONTENTS

The Alphabetical Index, pages 5-9, gives subjects in detail.

	PAGE		PAGE
WATER	11	COMPRESSED NON-VIB'R.	
STEAM	16	AIR ENGINE	166
COMBUSTION AND FIRING	19	MISCELLANEOUS questions	
Locomotive firing	29	and answers	169
BOILERS	33	Measurements and calculations	175
Rivets, Braces and Stays	33	MECHANICAL REFRIGERATION	180
Plant appliances	38	Ammonia	182
Boiler explosions	46	Methods of refrigeration	185
Running a boiler	49	Water examinations	195
Steam heating	52	Direct expansion system	197
Smoke and chimney	53	Ammonia tests, etc.	211
Brickwork	54	Review	219
Boiler testing	56	LIQUID AIR	223
Boiler horse-power	57	Liquid hydrogen	225
Feed water heater	59	THE MACHINE SHOP	226
Tensile strength	60	The lathe	228
SAFETY VALVES	62	Twist drill grinding	231
INJECTORS	69	Polygonal nuts	233
FEED PUMPS	73	RULES AND STANDARD NUMBERS	234
STEAM GAUGES	89	GENERAL USEFUL KNOWLEDGE	242
LUBRICATORS	91	ELECTRICITY	249
THE ENGINE	96	Dynamo and attachments	267
Valve setting	97	Varieties of dynamo	275
Reversing	100	Management	279
Lead and lap	102	Repairs	287
Compound engines	104	Measurements	290
Corliss electric engine stop	107	Motors	297
Hot air engine	109	Controllers	301
Condensers	112	Electric locomotive	304
The eccentric	116	Electric heating	306
Dead centers	118	Motor connections	308
Lining the engine	120	Electric wiring	309
Automatic governors	124	THE ELEMENTS OF ALGEBRA	318
Balanced slide valve	133	THE TRACTION ENGINE	324
Corliss engine and gear	135	THE HAY STACKER	332
Vacuum dash pot	140	JOURNAL BOX BABBITTING	335
Review	141		
Link motion	144		
HORSE POWER	146		
INDICATOR	154		
Pantograph	160		
Review	162		

INTRODUCTION

Almost every day some new device is invented for saving labor or fuel or other material. Where so many brains, scientifically trained, and so many thousands of practiced eyes and hands combine to make human life more comfortable, by shifting an ever larger portion of the hard labor to the shoulders of Nature's hidden forces, it is not strange that the engineer and machinist finds greater and greater demands made on his intelligence and experience.

A widely-known machinist delights in repeating to his friends his account of a little incident that will illustrate our point. He happened to enter the office of a large factory, where one of the firm jumped at him and hustled him into the engine-room, where the men tending the machinery were standing idle and puzzled. Something was wrong! "Start her up," said the proprietor. The big engine made two or three revolutions, giving a thump at each turn as if the fly wheel was about to go to pieces. "Stop her!" the machinist said, took the key of shaft and fly-wheel out, filed it down one sixty-fourth of an inch, and then drove it in place again—and she started up without a thump. "Well, I declare," said the proprietor, "how much do I owe you?" "Twenty-five dollars and fifty cents." "What's that, sir—$25.50 for twenty minutes of your time?" "No, sir; 50 cents for my time and $25 for knowing just what to do. It's worth that much to you, I dare say, to get your men to work, isn't it?" The money was cheerfully paid.

It's the PRACTICAL KNOWLEDGE that tells; and to aid engineers and mechanics in general to do intelligent work is the desire and aim of

THE AUTHOR.

LIST OF ILLUSTRATIONS

	PAGE
Cross Compound Engine..	10
Ice Plant	10
Smoke prevention	25
Riveting	33, 34, 35
Gusset stay	36
Steam fittings.	39
Globe valve..	40
Water column	42
Safety alarm	43
Boiler water line	45
Straightway valve	50
Boiler setting	54, 55
Safety Pop and Muffler	63, 64
Safety valve (lever)	68
Injector	70
Duplex pump	76
Check and gate valves	82, 83
Artesian pump	84
Pump governor	85
Deep well pump and plunger	88
Steam Gauge	89
Double and Triple feed lubricators	91, 93
Common slide valve and movements	96, 98, 100
Tandem Comp'd Engine	106
Corliss electric engine stop	108
Hot air engine	110
Connecting rod and oilers	111, 112
Concentric and eccentric	116
Direct and indirect motion	117
Slotted stick	121
Engine lining	122
Automatic governors	125, 130
Balanced slide valve	133, 134
Corliss valve movements	135, 136, 137
Corliss cut-off gear	138
Vacuum dash pot	140
Link motion	144
Indicator	154, 155
Indicator chart and diagrams	156, 158, 159, 160
Pantograph	161
Air engine	167
Crank pin travel	169, 170
Dead center points	172

	PAGE
Ice machine	184, 186
Freezing tank	192
Ice can dumps	194
Compressor valve	197
Gas compressors	199, 201
Ammonia liquefier	203
Valves and fittings	207
Bye pass valve	208
Ammonia testing	211
Engineer's and machinist's tools	226, 227
Lathe and tools	228, 230
Twist drill grinding	231, 232
Area of circle	234
Ventilation	243
Arc lamp	255
Incandescent lamp	261
Dynamo	268
Storage batteries	271, 272
Rheostat	272
Transformer and brush pointer	273
Alternating dynamo	276
Generator panel	279
Feeder panel	280
Chemical meter	295
Stationary motor	297
Carbon Brush Holders	298
Third rail motor trucks	299
Street car motor suspension and motor truck	300, 301
Electric car controllers	302
Electric locomotive	304
Electric Heating and Cooking	307
Stationary motor connections	308
Wire joints	314
Traction engine	324
Coal and water tank	325
Curve turning device	325
Compensating gear	326
Friction clutch fly wheel	326
Cross head	327
Tandem compound cylinders	329
Single eccentric reversing	330
Reversing rack	331
Hay stacker gearing	332, 333
Hay stacker	334

ALPHABETICAL INDEX

NOTE: For Electrical Terms see, also, Dictionary, page 253.

	PAGE		PAGE
Absorption method, ice	185	Boilers, Battery of	51
Accidents by shafting	245	Boiling, Definition of	18
Accumulator	271	Boiling points of ammonia	210
Air, a compound	13	Braces	35
" chamber, Duty of	83	Brine solutions, Table of	195
" Compressed—engine	166	" system	190
" in combustion	19	Brushes, Motor	298
" Liquid	223	Brush holder	273
" needed for fire	23	B. T. U.	293
" spaces in grates	20	Bye pass valve	207
" Weight of	13		
Alcohol expansion	15	Calculations, Engine	175
Algebra, Elements of	315	" of coal in bin	248
Alternating current	252	" Pulley speed	239
Ammonia	182	" Stay and bolt	38
" Boiling points of	210	Capacity of pump	74
" Charging	217	Carbonic acid	13
" compressors	184	Care of electric plant	284
" condenser	188, 202	Casing for electric wire	316
" Discharging	219	Cell, Daniell	274
" pump valve	197	" Gravity	274
" tests	211	Cement for steam pipes	222
" valve and fittings	207	Charging ice machine	217
Ampere	251	Check valve	81, 83
Ampere's Rule	252	Chimney	53
Appliances of steam plant	28	Circuit, Arrangement of	314
Area of circles	235	" Short	312
Artesian pump	84	Cleaning by steam	247
Atmosphere, Weight of	14	" belts	248
Automatic gov., slide crank	124	" rusty steel	247
" self-contained	129	Clearance	175
		Closed coil	276
Babbitting a journal	335	Clutch, Friction	326
Balanced slide valve	133	Coal, Decomposition of	21
Ball turning	229	" How it burns	22
Band saw mending	168	Coal-bin calculations	248
Barometer	14	Cold storage temperature	206
Battery, Electric	271	Color of flames	20
" of boilers	51	Combustion, Perfect	19
Belting, Cleaning of	248	Commutator	270
" horse power	152	Compensating Gear	325
Boiler, The	33	Compound engines	104, 328
" construction	45	Compressed air engine	166
" explosions	42	Compression method, ice	187
" How to clean	45	Compressor, Ammonia	197
" safe—pressure	176	Condensation, Ammonia	188
" setting	49, 54	Condensers, Ammonia	188, 202
" testing	56	" Jet	114

INDEX

Condensers, Open-air..... 204
" Steam....... 112
" Surface..... 113
Conductivity... 310
Connecting rod............ 122
Connections, Electric...... 314
Constant potential service. 278
Continuous current....... 252
Controller................. 301
Converter................. 273
Cooking, Electric 306
Corliss electric stop....... 107
" Engine 135
Coulomb.................. 291
Crankpin and crosshead
travel................. 169
Crosshead at dead center.. 172
" of engine 327
Current, Alternating..... 252
" Continuous..... 252
" Multiphase 252
Cut-out, Electric......... 316
Cylinder dimensions...... 328

Daniell cell............... 274
Dash pot.................. 141
Dead center............... 118
Deep well plunger........ 87
" " pump........ 87
Diagram, Indicator 156
Dictionary, Electrical..... 253
Differential gear 325
Dimensions of cylinders.. 328
Direct expansion system... 190
Discharging Ammonia
Pump 219
Distribution, Electric..... 314
Duplex gauge 90
Dyne 290
Dynamo and its parts 267
" Care of........... 279
" Efficiency of..... 293
" Repairs of........ 287
" Running a....... 283
" Varieties of. 275

Eccentric................. 116
" How to set an... 117
" rod............ 97
" Single-revers-
ing 100, 330
Efficiency of Dynamo..... 293
Electric locomotive........ 304
" heating and cook-
ing............. 306
" measurements.... 290

Electric wiring 309
Electricity, Chemical and
thermal..... 251
" Current and
statical....... 250
" Frictional and
voltaic 250
" Positive and
negative.... 249
Elements of Algebra...... 318
Engine, The 96
" Compound.. 104
" with
single valve.... 328
" Compressed air . 166
" Corliss.. 135
" stop, Corliss elec. 107
" Cross compound 105
" crosshead.. 327
" Electric 304
" Hot-air pump'g. 109
" Lining........... 120
" measurements . . 175
" pounding.. 119
" Receiver... 105
" striking points.. 122
" Tandem com-
pound 105, 328
" Traction......... 324
Erg 290
Expansion, Ammonia..... 188
" system, Direct 190
Explosion of boilers....... 46

Feed of boilers............ 43
" regulation........... 57
Fire, Care of.............. 25
' engine.... 83
Firing...... ...._..... 19
" Locomotive......... 29
" Stages of........... 20
Fittings, Heater and boil'r 39
Flames, Color of.......... 20
" Length of.......... 22
Foaming. ..... .'....... 50
Forced draught.. 22
Friction clutch 326
Friction in water pipe.... 238
Fuse, Safety, Boiler......42, 68
" " Electric..303, 308

Gaskets................... 51
Gas meter reading........ 243
Gauge, Comp'd ammonia. 90
" Duplex 90
" Steam 89

INDEX

Gauge, Vacuum........80, 115
Gear, Differential.......... 325
"   Reversing........... 330
Gearing, Lathe............ 228
"   Stacker........... 332
Glass tube, How to cut.... 247
Governor, Automatic, side
            crank..... ... 124
"   Autom., self-
        contained 129
"   Pump, Autom.. 85
Graphite for steam-fitting. 246
Grate, Air spaces in...... 20
Gravity cell.............. 274
Ground................... 312

Hard Water............... 195
Hay stacker............... 334
Heat...................... 174
"   Latent............18, 175
"   Utilized............. 23
Heater, Feed Water........ 59
Heating, Electric.......... 306
Horse power............... 146
"   Belting...... 152
"   Boiler....... 150
"   Compound
        engine.... 150
"   Electric..... 292
"   Evaporat'n.. 57
"   for incand.
        lamp 293
"   Heating sur-
        face.....37, 147
"   of traction
        engines... 328
"   of waterfall. 147
"   of water-
        wheel.... 148
"   Rating of... 149
"   Steam con-
        sumption. 149
"   Tubular
        boiler..... 152
House Wiring............. 309
Hydrogen, Liquid......... 225

Ice Making............... 180
Incand. Lamp, H. P. for... 293
Indicator................. 154
"   card............. 153
"   diagram chart ... 158
"   examination..... 162
"   with pantograph. 160
Induction................. 250
Injectors, Classes of....... 71

Injectors, Parts of 70
"   Size of........... 72
"   Working of..... 69
Insulation................ 311
"   testing......... 281
Inverse ratio 310
Iron and steel..........61, 128

Joints, Electric 314
Journal babbitting........ 335

Kilowatt 148, 264

Lamp sockets............. 317
Latent heat...........18, 175
Lathe gearing............. 228
"   tools............. 230
Law of Ohm............. 291
Lap and lead............ 102
Leather belting cleaned. 248
"   H. P...... 152
Leverage 241
"   in safety valves.. 66
Lightning, What is........ 251
Lining an engine.......... 120
Link motion.............. 144
Liquid air................. 223
"   hydrogen......... 225
Lubricators, How to attach 92
"   How to clean. 85
"   How to run .. 93
"   Triple sight.. 93
"   Working of .. 94

Machine shop............. 228
Magnetic field............ 250
Measures and weights..... 236
Measurements, Engine.... 175
"   Electric..... 290
"   Chemical. 293
"   Mechani-
        cal...... 286
Mending band saw........ 168
Meter, Gas............... 243
"   Chemical 293
"   Mechanical........ 286
Mineral water............. 196
Miner's inch 258
Miscellaneous Q and A.... 169
Molecules 171
Motions, Direct and Indi-
            rect.......... 117
"   of Crosshead..168, 172
Motors, Stationary 296, 308
"   Third rail......... 298
"   Surface Road..... 300

INDEX

Motor Brushes. 273, 298
" Reversing 308
Muffler, Safety Valve...... 63
Multiphase current...... 252
Multipliers, Standard..... 235

Nitrogen 13
" Weight of. 19
Nuts, Polygonal 233

Ohm, Law of............... 291
Open coil.................. 277
Over and under........... 98
Oxygen..................... 13
" Weight of........ 19
" in Combustion... 19

Pantograph 160
Parallel connection 277
" wiring........... 315
Pipes, Standard Threads on 39
Placing Elect. Wires...... 315
Plant, Running Elect. 284
Plunger, Deep Well....... 87
Pole, Positive.............. 252
Polygonal nuts............ 233
Potential.................. 251
" service, Constant 278
Pounding in engine 119
Pressure in stand pipe.... 14
" Initial........... 102
" Safe boiler... 176
" Terminal. 102
Priming 51
Pulley speed calculation .. 239
Pump, Ammonia........... 197
" Artesian............ 84
" Capacity of........ 74
" Deep well......... 87
" Duplex v'lv. setting 75
" Feed............... 73
" governor........... 85
" Lift of............. 81
" testing............. 79

Quadrant.................. 273

Rack, Reversing 331
Recipe, mending band saw 168
" Test'g iron and steel 128
" Tracing paper...... 274
" Solder and fluid.... 205
" Steam cement...... 222
Refrigeration............. 180
" Apparatus. 190, 197
" Methods of... 185

Refrigeration, Principle of. 181
" Testing ice machinery. 214
Regulation of Feed........ 57
Repairs of Dynamo....... 287
" Band Saw...... 168
Resistance in Controller .. 302
" of metals...... 310
Rheostat.......... .....272, 308
Rivets 33
Reversing rack............ 331
" an engine...... 100
" a motor......... 308
Rubber gloves, Use of ... 289
" belting, Cleaning. 248
Reversing gear, Single eccentric. 330
Rules and standard numbers................... 234
Running electric plant.... 284
Ruptures of boiler....... 56
Rusty steel, Cleaning...... 247

Safety fuse, Boiler........42, 68
" " Electric...303, 308
" pop valve........... 62
" valve, Setting...... 65
" " Size of........ 64
Safe working pressure.... 48
Series wound 277, 308
Shafting accidents 245
Shaft lining, Engine....... 120
Short circuit.............. 312
Shunt wound.............. 277
Slide valve................ 96
" " Balanced........ 133
" " reversing....... 100
" " setting......... 97
Smoke.. .., ...21, 53
" prevention......... 25
Sockets, Electric.......... 317
Solder and fluid........... 205
Specific gravity........... 13
" " of ammonia 213
Square root............... 239
Stacker, Hay.............. 334
" gearing........... 332
Standard babbitts......... 335
" multipliers...... 235
" numbers......... 234
" threads on pipes 39
Stationary motor.......297, 308
Stays, bolts, calculation...37, 38
Steam 16
" Cleaning by.......... 247
" expansion 18

INDEX

Steam fitting............ 246
" gauge 89
" " test........... 65
" heating............. 52
" High, low......... .. 17
" pipes, Cement for... 282
" pressure and temp. 17
" Superheated......... 17
" velocity 18
Steel and iron............61, 128
Striking points 122
Substance, Three forms of. 11
Surface car, Elect.......... 300
Synchronizing............. 278

Table, Ammonia per cent.. 213
" " boil. p'nts 210
" Areas of circles..... 286
" Boiler H. P......... 153
" Brine solutions..... 166
" Cold storage... 206
" Conductivity of
    metals............ 310
" Engine H.P........ 149
" Flame temperature. 21
" Grate space........ 20
" Heat'g surface H.P. 118
" Indicator springs... 174
" Polygonal nuts 283
" Rivet sizes. 35
" Steam pressure..... 17
" " velocity..... 18
" Standard babbitts .. 335
" Standard threads... 39
" Traction engine H.P. 328
Tensile strength........... 60
Testing ammonia.......... 211
" boilers........... 56
" circuit........... 284
" ice machinery..... 214
" insulation......... 281
" iron and steel...... 128
" pump............. 79
" steam gauge....... 65
" water............15, 195
Thermal unit............. 23
Thermometers 175, 244
Third rail system......... 300
Threads of pipe, Standard. 39
Tracing paper, Recipe for, 234

Traction engine.......... 324
" " H. P....... 328
Transformer...... . ....272, 293
Travel of crankpin and
  crosshead................ 169
Turning a ball............. 229
Twist drill grinding. 231

Unit, Electrical........... 266
" Thermal............. 23

Vacuum, Ammonia pump. 216
" Water pump...81, 115
Valve, Bevel of... 66
" Bye pass.......... 297
" Check 81
" Corliss........... 135
" Duplex, To set..... 75
" Gate.............. 82
" Link motion....... 144
" rod, Length of..... 97
" rod adjusting, Cor-
    liss............. 139
" Safety 82
" Slide.............. 95
" Slide, Balanced.... 133
Ventilator............... 242
Vibration, Anvil........... 246
Volt..................... 292

Water.................... 11
" Boiling points of... 16
" column............. 42
" Composition of..... 12
" expansion........... 12
" Freezing........... 180
" Purifying........... 12
" measurements...... 15
" as a solvent 12
" tests.............15, 195
Watt..................... 290
Weights and measures.... 296
Wire, Placing............. 315
" Size of............?.. 313
Wiring, Electric........... 309

Yoke and Quadrant....... 273

Zero, Absolute............. 174

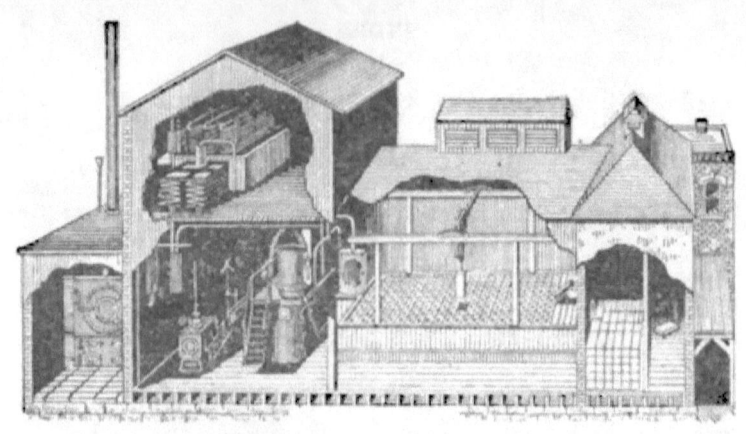

ICE MANUFACTURING PLANT.

CORLISS CROSS COMPOUND ENGINE.

Stevens' Mechanical Catechism

WATER

Life, as it exists on our earth, depends on water and heat. Water is the most important substance in nature.

QUESTION.—How does life depend on water?

ANSWER.—Water is present everywhere, in the air, in the ground, in wood and even the hardest stone. About seven-eighths of the human body is water. Without water everything would be dry and lifeless.

Q.—What are the most important qualities of water?

A.—First, its abundance and universal presence; second, its quality of assuming easily either of the three forms of substance, *solid*, *liquid* and *gaseous*. Many substances can be in these three forms, but water changes from either one of them to the others within a very narrow range of temperature. It freezes solid (ice) at 32° F. (=0° C.) and turns to gas (vapor) at any temperature, most rapidly at the boiling point (212° F. or 100° C.)

Q.—Is water an element or a compound?

A.—A compound, composed of two gases, hydrogen and oxygen, in the proportion of one volume of oxygen to two volumes of hydrogen, or in *weight* one part of hydrogen to eight parts of oxygen.

Q.—Can water be condensed by pressure?

A.—Very slightly. Under a pressure of one atmosphere it may be compressed only about one twenty-thousandth of its bulk.

Q.—Has water any solvent power?

A.—Yes, it is the most universal and powerful solvent of all liquids. For this reason it is rarely found entirely pure.

Q.—How can water be entirely purified?

A.—By changing it to steam and condensing this.

Q.—What taste or color has pure water?

A.—Pure water is tasteless, odorless, colorless and transparent.

Q.—Does water expand or contract when freezing?

A.—It expands about 1-12 of its bulk.

Q.—When has water the smallest bulk?

A.—At the temperature of 39.1° F. (=4° C.)

Q.—Would you call the expansion of water in freezing a force?

A.—Yes. It exerts the tremendous power of 30,000 lbs. per sq. inch.

Q.—What is *specific gravity?*

A.—Density as compared with water. 92 lbs. of ice equal 100 lbs. of water at 60° F. in volume.

Q.—State the average impurities of saline matter in the Atlantic Ocean and in the Dead Sea?

A.—The saline matter in the Atlantic Ocean amounts to 2,139 grains per gallon, and in the Dead Sea it reaches 19,736 grains per gallon.

Q.—What is the proportion in area of land and water on the globe?

A.—About 52 millions sq. miles are land, and about 196 millions sq. miles are water.

Q.—Where do the clouds come from?

A.—They are formed by the constant evaporation from the immense water surfaces of the globe.

Q.—Is air an element or a compound?

A.—Like water, it is a compound, composed of 20.96 per cent oxygen, 79.00 nitrogen and 0.04 per cent carbonic acid.

Q.—Which of these gases is the life-sustaining element?

A.—It is the oxygen. This is the substance whose chemical union with combustibles we call combustion, whether in our lungs or in a fire-box.

Q.—What is the difference in weight between air and water?

A.—Air is 813.67 times lighter than water.

Q.—Can it be proved that air has weight?

A.—Yes, by comparing the weight of a large

hollow globe when filled with air, with its weight after the air has been exhausted by an air pump.

Q.—How much weight has the atmosphere per sq. inch?

A.—The mean pressure of the atmosphere is stated at 14.7 lbs. per sq. inch.

Q.—How is it that this weight does not crush us?

A.—The pressure is exerted in all directions, and permeates the whole body.

Q.—What is the meaning of such terms, as two or three "atmospheres"?

A.—An **atmosphere** in this sense is the standard or unit of air pressure, equal to the average atmospheric pressure at sea level (=14.7 lbs. per sq. inch).

Q.—Is the atmospheric pressure not always the same?

A.—No. The barometer shows the variations.

Q.—What is the principle of the barometer?

A.—The mercury in the closed vacuum tube is raised about 29.9 inches by the atmospheric pressure entering through the open tube.

Q.—How high will the atmospheric pressure raise water in a vacuum tube?

A.—About 33.9 feet.

Q.—What is the pressure in pounds per sq. inch in a column of water in a standpipe?

A.—Multiply the height of the column in feet by .434. Engineers generally figure one-half of one pound pressure per sq. inch for each foot elevation.

Q.—What are the common measures of weight and contents for fresh water?

A.—A gallon weighs 8 1-3 pounds and contains 231 cubic in. A cubic foot weighs 62 1-2 pounds and contains 1,728 cubic in., or 7 1-2 gallons.

Q.—What kind of water would you prefer to use in a boiler?

A.—Rain or atmospheric water.

Q.—Why do you prefer rain water?

A.—Because it does not contain minerals which scale the boiler heavily.

Q.—Is rain water considered pure?

A.—Yes, but in their fall raindrops collect many solid particles of dust, both in the air and on the ground. On the other hand, spring water contains almost invariably mineral matter, which causes corrosion and slight deposit in the boiler. Rain water is almost entirely free from elements that cause incrustation.

Q.—What tests are there for impure water?

A.—Litmus paper dipped in vinegar does not return to its true color in water containing earthy matter or alkali. A solution of a little prussiate of potash will turn water containing iron blue. A few drops of a solution of a little good soap in alcohol, if put in a vessel of water, will turn it quite milky if it is hard; soft water will remain clear.

STEAM

Many engineers have asked for an explanation of the term "Steam," which we have endeavored to give in the following questions and answers:

Q.—What is "Steam"?

A.—Steam is the gas from water produced by ebullition, which is generally known to take place at 213° F. The passage of any liquid into the gaseous state is called vaporization, and the term "evaporation" especially refers to the slow production of vapor at the free surface of a liquid. In *boiling* vaporization goes on not only on the surface, but in the liquid itself.

Q.—Is the boiling point of water under all circumstances at 212.8° F.?

A.—No. On high mountains, where the atmospheric pressure is very low, water boils at a much lower temperature, so that cooking cannot be done except in air-tight vessels; and under high pressure, as in steam boilers, water begins to boil at a much higher temperature.

Q.—State if the temperature of the boiling point of water increases the same as the steam pressure?

A.—No. The following table will explain the different temperatures at different pressures:

STEAM PRESSURES AND TEMPERATURES

Pressure.	Temp.	Pressure.	Temp.
10	192.4	75	311.0
15	212.8	80	315.8
20	228.5	85	320.1
25	241.0	90	324.3
30	251.6	95	328.2
35	260.9	100	332.0
40	269.1	120	345.8
45	276.4	130	352.1
50	283.2	140	357.9
55	289.3	150	363.4
60	295.6	160	368.7
65	301.3	170	373.6
70	306.4	180	378.4

Q.—What is the difference between high and low pressure steam?

A.—High pressure is steam over 15 lbs.; low pressure is 15 lbs. or less.

Q.—What is superheated steam?

A.—Steam removed from the water boiler and brought to higher temperature in a separate vessel.

Q.—Does water evaporate when the air above its surface is exhausted by an air-pump?

A.—The temperature could be elevated to 275° before vaporization takes place, and when it does, the action will not be like ordinary ebullition under pressure of the atmosphere, but will be instantaneous (explosive).

Q.—What is the boiling point of a liquid?

A.—A liquid boils when the tension of its vapor and the pressure it supports are equal.

Q.—What is meant by "latent heat" in connection with steam?

A.—The effects of heat upon a body are, 1, increase of temperature; 2, expansion, or increase of volume; 3, change of state, as of a solid to a liquid, or of a liquid to a gas. To transform water at 100° C. into steam, a large amount of heat is required, which disappears as sensible heat, and is said to become latent.

Q.—Which has the more expansive force, water or alcohol?

A.—Alcohol has more than double the expansive force of water of the same temperature. The steam of alcohol at 174° is equal to that of water at 212°. When proper means can be invented for saving the fluid from being lost it is supposed that alcohol can be employed with great advantage as the moving power of engines.

This table shows the velocity (per second) at which steam escapes into the atmosphere at given lbs. of pressure above one atmosphere.

PRESSURE IN LBS.	VELOCITY IN FEET.	PRESSURE IN LBS.	VELOCITY IN FEET.
1	540	50	1736
3	814	60	1777
5	981	70	1810
10	1232	80	1835
30	1601	100	1875
40	1681	120	1900

COMBUSTION AND FIRING

Q.—What is meant by combustion?

A.—It is a chemical combination of oxygen with combustible material of any kind, commonly called fire.

Q.—State the comparative weight of oxygen to nitrogen?

A.—It is 16 to 14.

Q.—How much air is necessary to consume a given quantity of fuel?

A.—There is a fixed proportion between the oxygen required and the fuel gas to be consumed.

Q.—How can this be determined?

A.—We know that oxygen is 1-5 the bulk of air. Five volumes of air are necessary to produce one of oxygen; therefore, as two volumes of oxygen for each of gas are necessary, it follows we must provide ten volumes of air.

Q.—What is perfect combustion?

A.—Combustion is perfect, when no gas is developed that does not instantly unite with oxygen.

Q.—What amount of air is required to consume 1 lb. of coal?

A.—It requires 15 lbs. of air, on an average.

Q.—How many cubic feet in 1 lb. of air?

A.—One lb. of air contains 13 9-107 cubic feet

Q.—Give the proper air spaces for different fuels?

A.—The different **sizes of air spaces** between grate bars for different fuels are as follows:

Schuylkill anthracite pea coal............	¼ inch
Lehigh anthracite pea coal............	⅜ "
" " chestnut............	⅜ "
" " stove............	½ "
" " broken............	⅝ "
Cumberland bituminous............	¾ "
Wood............	¾ to 1 "
Sawdust............	3-16 to ¼ "

Q.—Of what **color** are **flames** in different temperatures?

A.—They are as follows:

Color.	Temp.	Color.	Temp.
Red, just visible..	977 deg.	Orange, deep.....	2010 deg.
" dull.........	1290 "	" clear....	2190 "
" cherry, dull..	1470 "	White heat.......	2370 "
" " full..	1560 "	" bright......	2550 "
" " clear.	1830 "	" dazzling...	2730 "

Q.—State the successive stages of firing coal, and the temperatures, beginning with the match?

A.—A slight friction of 150° ignites the phosphorous; when this reaches 500° the sulphur burns; next 800° ignites the wood or shavings, then 1,000° ignites the coal gas.

Q.—How is the philosophy of combustion known?

A.—It is known through chemistry.

COMBUSTION AND FIRING

Q.—What do you understand by coal?

A.—Coal is a compound substance and may be decomposed by heat in several distinct elements.

Q.—Which elements are of principal importance in the combustion of coal?

A.—Two: carbon in form of coke, and hydrogen, a gas.

Q.—Are these all the heating properties in coal?

A.—No, but the principal ones.

Q.—Why does not coal commence to burn immediately when thrown upon the fire?

A.—Because coal must first be heated and go through the process of decomposition.

Q.—**How does coal decompose in firing?**

A.—100 lbs. of coal, when put in a fire, develops gases containing about 24 lbs. of hydro-carbon and free hydrogen, 9 lbs. of steam (water), 1.25 lbs. of sulphur and 1.2 lb. of nitrogen. While these are evaporated and consumed, the residuum, about 60 lbs. of fired carbon or coke, begins to burn, leaving finally about 4.55 lbs. of ashes (incombustible matter).

Q.—What condition does smoke indicate?

A.—It indicates poor combustion.

Q.—Is it understood, then, that when there is no smoke there is perfect combustion?

A.—No. The perfect combustion of coal in a furnace can only be effected by a large enough supply of **oxygen.**

Q.—What is the object of a forced draught?

A.—To increase the supply of oxygen.

Q.—Does coke smoke while burning?

A.—No. Smoke only comes from the gas distilled from the coal. After the gas is distilled, that which is left is the coke.

Q.—Where is the greatest heat when gas is being expelled from coal?

A.—In the gas.

Q.—Will the coke or solid coal burn while expelling gas?

A.—No. A lump of coal may, however, be expelling gas in one place and burn in another where the gas has already been expelled.

Q.—What is required to burn anything?

A.—In order to burn anything it must be heated to a certain degree and kept up to that heat.

Q.—**How far does a flame enter** a boiler tube of ordinary size?

A.—The flame never enters more than a few inches.

Q.—Then state what burns at the other end of the tube?

A.—It is carbonic oxide. It has a low igniting temperature and takes fire after mixing with the atmosphere, making a blue flame attending the conversion of carbonic oxide into carbonic acid.

Q.—The blue flame just spoken of—is it the same that entered the tube?

A.—No. The flame that entered the tube was extinguished, and any combustible matter still present went to waste.

Q.—1. What is a thermal unit? 2. How many does a pound of good coal yield?

A.—1. It is the heat required to raise the temperature of 1 lb. of water 1° F. 2. 13,000.

Q.—What becomes of the heat?

A.—Fifty per cent is utilized, 40 per cent escapes up the chimney, and 10 per cent is lost by radiation.

Q.—State the weight of 1 cubic foot of air?

A.—A cubic foot of air weighs 535 grains.

Q.—State the amount of air required for the combustion of 1 lb. of good free burning soft coal?

A.—About 200 cubic feet.

Q.—Is combustion always accompanied by flames?

A.—No. **Combustion is, chemically, a rapid oxidation**, caused by the chemical union of oxygen with the combustible. The rotting of vegetable matter, the rusting of iron, the oxidation of brass are examples of combustion without flame. They cannot take place without air, which furnishes the oxygen. Combustion takes place in our lungs, which absorb the oxygen. *Pure, fresh air contains oxygen in abundance.

Q.—Where do we get the heat from?

A.—The heat is produced by the chemical union of the air with the carbon and hydro-carbon of the fuel. We might as well expect to make steam without putting fuel on the grates as without supplying the fuel with the proper amount of oxygen as contained in air.

Q.—About how much air enters a furnace having a good natural draught?

A.—About 530 cubic feet per minute to each square foot of grate surface.

Q.—Suppose you had a 60 H. P. boiler with 25 sq. feet of grate surface, allowing about 25 per cent of the grate surface for air space, how many square feet for air would there be, and how much would pass through the grates; also how many pounds of coal would be consumed per hour?

A.—The air space would be $6\frac{1}{4}$ sq. feet, the amount of air passed through would be about 198,720 feet and the amount of coal about 1,000 lbs. per hour. In seven cases out of ten it will be found that the grates are choked with clinkers and the ash-pit filled with ashes, so that not more than 25 per cent of stated amount of air could possibly reach the fuel. The fireman shovels in coal and wonders why he can't raise the steam pressure, never dreaming that the required amount of air for the combustion of the amount of fuel thrown in could not possibly pass through the dampers, **much** less through the clogged grates.

Q.—Should the grate slope, and, if so, in which direction and how much?

A.—It should slope ¼ inch per foot, from the front of the furnace to the bridge wall, downward.

Q.—What fireman will best prevent the offensive and wasteful formation of smoke?

A.—A fireman who keeps the grates free for the passage of air and always breaks up the coal into lumps about the size of a man's fist and keeps it evenly distributed over the grates.

Q.—Do you know of any practical and cheap **device for the prevention of smoke?**

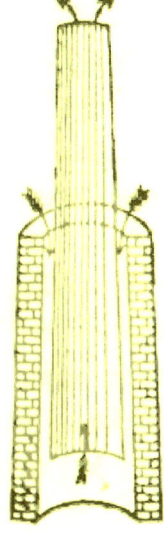

A.—A pipe inserted in the top of the stack, leaving an annular space of three inches between itself and the inside of the stack, and extending about eight feet down into the stack and projecting seven feet above it, is said to be an excellent and cheap device for preventing smoke and to save ten per cent of fuel besides.

Q.—How does it work?

A.—At starting the fire, thick smoke issues from the inserted pipe, while faint rays of smoke issue from the annular space. These almost immediately disappear and from the central pipe only traces of smoke can be seen to issue. This proves that a cold circular draught descends around the hot upward draught, and reaching the combustion chamber hot and in abundance, improves the combustion.

Q.—When should fresh coal be thrown in on the fire?

A.—When the whole fire has reached a white heat, the door may be opened and a few shovelfuls of coal thrown on the front of the grates and the door closed as quickly as possible.

Q.—What does this do?

A.—The gases distilled from the fresh coal will be ignited while passing over the hot coals on the rear of the grates, and instead of giving off a dense black smoke the hydro-carbons will be entirely consumed.

Q.—What should be done then?

A.—When the coal has ignited it may be pushed back over the grates and a fresh supply thrown in front again. This kind of firing will prevent smoke, if anything will, but it is hard work for the fireman, and when such services are given they will be appreciated and encouraged by the employer.

Q.—About how thick should fires be for different coals?

A.—For anthracite coal the thickness should be about 8 inches, for soft coal about 10 inches and for coke about 12 inches.

Q.—Suppose you could not carry a fire bed of the desired thickness without blowing off steam, what would you do?

A.—I should reduce the grate surface area by laying in fire brick next to the bridge wall and next to the sides of the furnace to the height of 8 or 10 inches.

Q.—In starting a fresh fire under a cold boiler how would you proceed?

A.—First, have two gauges of water in the boiler, then cover the grate bars all over with coal, leaving a space in front for some light wood and shavings; cover the back with some heavy wood on the coal, close the ash-pit doors tightly and partly close the furnace doors when the wood is lit. The coal on the grate prevents warping.

Q.—Why not place the coal on top at the beginning?

A.—Because it would prevent the free access of air to the wood. The air enters through the furnace doors and the draught carries the flames between and over the coal in the rear, gradually heating it and distilling the gases out of it, which ignite, adding to the heat.

Q.—Then what should be done?

A.—After wood is burning, coal should be thrown on, the furnace doors closed and pit doors opened.

Q.—Is it a good idea to hurry a fire?

A.—No. It should be allowed to burn gradually by feeding the fire with a little coal at a time.

Q.—Is it good policy to stir a fire often?

A.—No. It should be left alone as much as possible.

Q.—Why is it not good policy to stir a fire often?

A.—Because it would have a tendency to drop all the small coal and fire through the grate.

Q.—How is the draught controlled?

A.—By the chimney damper and ash-pit doors.

Q.—How often should a fire be cleaned, and when?

A.—As often as necessary—when clinkers prevent the admission of air through the grates.

Q.—Can it be seen by the color of a fire when it should be cleaned or is badly managed?

A.—Yes. Dark spots, heavy smoke and blue flames are the best points to show it.

Q.—How would you clean a fire?

A.—Open one of the furnace doors, shove the live coals either back or to one side; then rake out the dead clinkers, throw in a little wood or coal and pull the live coals over; then throw on fresh coal.

Q.—Would you use wet coal?

A.—Some engineers do not believe wet coal harmful, but others do.

Q.—How and when would you bank a fire?

A.—First, clean the fire on one side of the furnace, throw on a few shovels of fine coal and cover it with wet ashes, tightly close the ash-pit and furnace doors, leaving the stack damper slightly open to let out the gas. A fire banked in this manner will keep all night.

Q.—How many gauges of water would you consider safe with a banked fire?

A.—Three full gauges.

Q.—Is it a good idea to entirely close the chimney damper with fire on the grates?

A.—No. It is dangerous, as gas may collect in the flues or tubes, and cause an explosion that might do very serious harm.

Q.—Does a banked fire benefit a boiler?

A.—Yes. It prevents any contraction owing to the difference in temperature.

LOCOMOTIVE FIRING

Q.—How would you fire a locomotive boiler on the road, run light on coal, avoid much smoke and have the boiler steam well?

A.—Fire a little at a time and often, also keep the fire level as near as possible. Fire with evensized coal and look out for clinkers in the box; also close door after each shovelful.

Q.—What understanding have you of steam pressure as shown on the gauge?

A.—It indicates the pressure on each square inch against the inside of the boiler.

Q.—Where would you place a safety plug in a boiler having a fire-box?

A.—In the center of the crown sheet.

Q.—Explain why **steam is exhausted through the stack?**

A.—Without it the draught would be too weak for the needs of a locomotive. It forces the air out of the front end up the stack, creating a draught which causes the gases and products of combustion in the fire-box to fill the space; this in turn allows the pressure of the atmosphere to force fresh air up through the grates, making a steady and strong flow of air into the fire-box.

Q.—Is enough air supplied through the grates to form perfect combustion?

A.—Not under all circumstances.

Q.—Are there other ways of admitting air into the fire-box?

A.—Yes. Air is admitted over the fire through hollow stay-bolts, also air holes in the fire-box door and lining.

Q.—Do the holes in the door answer any other purpose?

A.—Yes. If drilled in line with the lining holes, the light from the fire will light up the deck and coal space.

Q.—Does the cold air that is admitted over the fire mix with the gas and burn immediately upon entering the fire-box?

A.—No. It is heated first, then mixes with the gas and burns.

Q.—State the object of the brick arch in the fire-box?

A.—It is there to hold the gas expelled by the coal so it will mix with the air admitted. It heats the air and prevents the emission of dense black smoke. It protects the flues from the cold air that passes through the door when firing, and checks the exhaust's effect upon the fire, so that small particles of coal that would otherwise go through the flues and be lost, are kept in the fire to be burned.

Q.—Is the brick arch a coal saver?

A.—Yes. It saves coal by holding in the gas so it can burn, and prevents the flue sheets and flues from sudden cooling when the fire-box door is opened. An arch is a disadvantage if the side sheets are patched or leaking, as the arch makes them worse. It keeps them hot after the other parts of the fire-box are cool, consequently it causes expansion where there should be contraction.

Q.—What effect does an open door have on the fire and flue sheet when an engine is working?

A.—It lets the air in through the door instead of through the fire, which cools the flue sheet and lowers the pressure. When firing see that the door is closed after each scoop of coal.

Q.—Why open and close the door so often?

A.—It gives each previous shovelful a chance to ignite.

Q.—Is the **wetting of coal** for locomotives any advantage over dry coal?

A.—No, but with large lumps the water gets in the cracks and splits the lumps as soon as heated, and for small coal it helps to coke into a lump, so that it will stay in the box and burn instead of going out with the first exhaust.

Q.—Of what use is a **blower?**

A.—It is very useful. It makes a forced draught, which prevents black smoke, and keeps the smoke and fire in circulation when engine is shut off.

Q.—How is smoke kept from trailing over the train when running shut off?

A.—Sometimes partly opening the door will remedy the trouble, otherwise the blower must be turned on a little to force the draft. Good judgment should be used.

Q.—Is it wasteful to have an engine frequently blow off at safety valve?

A.—Yes; but if the pressure can be kept within 5 lbs. of the blowing off point it will be easier on the boiler and will save water and coal.

Q.—Give the proper **size of a locomotive stack** —inside diameter?

A.—It should be $2\frac{1}{2}$ inches **smaller in diameter** than the cylinder of the **engine.**

BOILERS

Locomotive and Stationary

Q.—State the different classes and styles of boilers in use?

A.—There are three classes—marine, stationary and locomotive—and six styles—marine, locomotive, upright, flue, tubular and water tube boilers.

Q.—How are the different classes fired?

A.—The marine, upright and locomotive (hanging fire-box) are fired internally, but the stationary boilers are mostly fired externally.

RIVETS

Q.—Are boiler shells single or double riveted?

A.—The end seams are all single riveted. The longitudinal seams are single riveted for low pressure and double for high pressure.

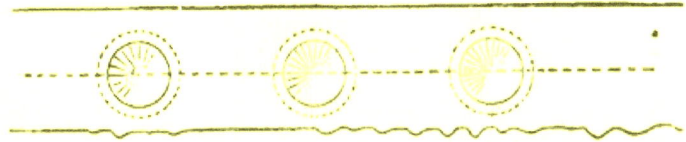

Q.—Why are longitudinal seams double riveted and circular or end seams single riveted?

A.—Because the strain is greater on the sides

than at the ends, as the steam pressure has more surface to work on.

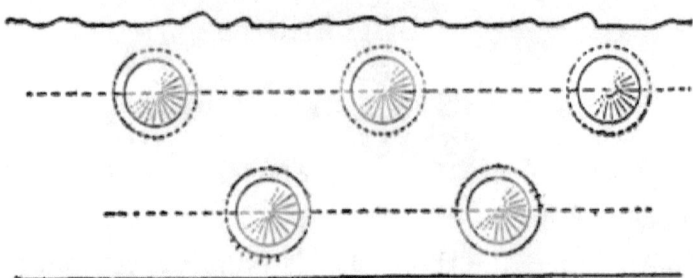

Q.—What is the distance generally between rivets of a single, or double riveted boiler shell?

A.—Single rivets are generally 2⅛ inches, and double rivets 2¾ inches apart.

Q.—What should the diameter of rivets be for any size sheet to make up the maximum shearing strength?

A.—The diameter should be equal to twice the thickness of plate to be riveted.

Q.—What is the usual distance between the edge of rivet hole and edge of sheet?

A.—The full thickness of rivet used.

LAP JOINT RIVETING

The following table indicates the various sizes, etc., of rivets for plates of different thickness:

SINGLE RIVETED LAP JOINT.

DOUBLE RIVETED LAP JOINT.

BOILERS

THICKNESS OF PLATE.	DIAMETER OF RIVET.	DIAMETER OF HOLE.	PITCH.		STRENGTH IN % OF SOLID.	
			SINGLE.	DOUBLE.	SINGLE.	DOUBLE.
1/4 in.	5/8 in.	1 1/8 in.	2 in.	2 1/4 in.	0.66	0.77
5/16 "	11/16 "	3/4 "	2 1/16 "	2 5/8 "	0.64	0.76
3/8 "	3/4 "	13/16 "	2 1/8 "	2 3/8 "	0.62	0.75
7/16 "	13/16 "	7/8 "	2 3/16 "	2 7/8 "	0.60	0.74
1/2 "	7/8 "	15/16 "	2 1/4 "	3 "	0.58	0.73

This table is applicable when steel rivets are used in steel plates, or iron rivets in iron-plates. When iron rivets are used in steel plates, both rivets and rivet holes should be larger by 1-16 of an inch.

SINGLE RIVETED BUTT JOINT.

DOUBLE RIVETED BUTT JOINT.

When plates thicker than 1/2 inch are used, the joint should be a butt joint with double fish plate. (See cuts.)

BRACES AND STAY BOLTS

Q.—Where should braces be put in a fire-box boiler?

A.—On the crown sheet, in water leg, in dome and on all flat surfaces.

Q.—What kind of braces should be used?

A.—In the dome, crow foot or solid braces; on flat surfaces and between water sheets, stay-bolts; on the crown sheet, crown bars and stay-bolts; in the boiler shell crown radial braces; and in corners gussets.

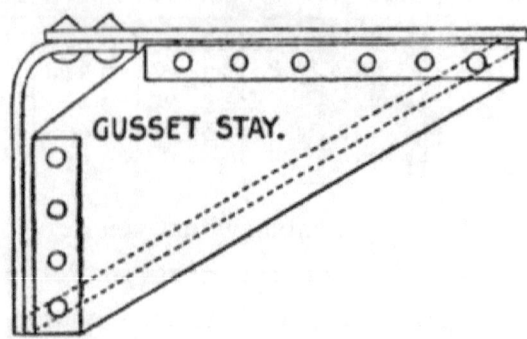

Q.—How is the load on a brace calculated?

A.—Multiply the supported area by the steam pressure and divide the product by the number of braces. The quotient gives the strain on each brace. The law allows not more than 6,000 lbs. per sq. inch of cross section of brace. A round brace of 1⅛ inch diameter has 1 sq. inch area in cross section.

Q.—How can it be known whether a brace is really carrying the intended load?

A.—If it does, it will give an even, clear sound when tapped with a hammer or the like.

Q.—How are braces put in properly?

A.—Have the brace about 1-16 of an inch short, heat it red hot in the center and put in place. It will shrink tight when it cools.

Q.—Why is the flue sheet thicker than the boiler shell sheets?

A.—Largely because it is weakened by the many

flue holes cut in it, and it has to support the weight and sag of the flues.

Q.—Give the number of square feet of heating surface allowed to a horse-power in different types of boilers?

A.—For vertical 12 sq. feet, for horizontal tubular 15 sq. feet.

Q.—How is a boiler's horse-power determined?

A.—Add together all the areas, in sq. feet, of heating surface up to the fire line (shell, tubes, back head); subtract from this sum the cross section area of all the tubes and the area of the front head less the tubes, and divide the remainder by 15 if horizontal, by 12 if vertical. See pages 148 and 153.

Q.—Give tonnage strain on the crown sheet of a fire-box?

A.—Multiply the length by breadth inches, divide by 12 for feet, multiply answer by steam gauge pressure and divide by 2,000.

STAY-BOLTS

Q.—Explain the use of stay-bolts?

A.—They are used to strengthen flat surfaces in steam boilers.

Q.—State the surface of plate a stay-bolt must support?

A.—The support is represented by the area enclosed between four bolts.

Q.—How is the area between the four bolts found?

A.—By multiplying one distance by the other. The answer will be each bolt's support.

Q.—What pressure do the four bolts have to withstand?

A.—Multiply the area by highest boiler pressure. The product is the strain on cross sectional area.

Q.—State the strain on a single stay-bolt?

A.—It must not be over 6,000 lbs. per sq. inch cross sectional area. Rule: Multiply cross sectional area of bolt by 6,000, divide by steam pressure and extract square root of quotient. (See under Miscellaneous, page 239.)

Q.—In examining the inside of the boiler, what are some of the defects for which you would be on the lookout?

A.—For missing pins from the braces, slack braces, leaky socket bolts, defective riveting, defective heads to the rivets and for broken or loose stays.

Q.—Name some **appliances necessary** about a steam plant?

A.—A boiler and fittings, a pump or injector, piping for the feed water apparatus, steam pipes, globe valves, feed valves, feed water heater, steam trap, chimney and dampers, safety valve, check valve, the fire front containing the fire and pit, also flue doors, grate bars, and bearing bars, dead plates, man and hand hole plates, thimbles, water gauge cocks and glass gauge, blow-out cock, fusible plugs, steam gauge, fire tools, flue brush, gaskets and scaling tools, also hose for washing out the boiler, shovel, slice bar, rake, hoe, etc.

STEAM FITTINGS

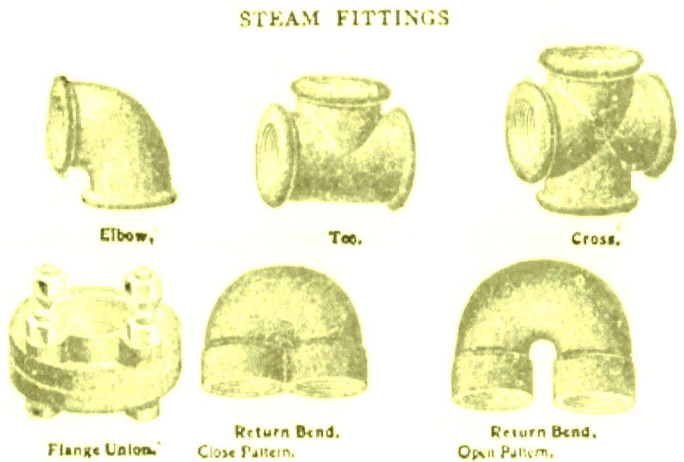

Elbow.   Tee.   Cross.

Flange Union.   Return Bend, Close Pattern.   Return Bend, Open Pattern.

Pipes of $\frac{1}{8}''$ bore have 27 threads to the inch; pipes of $\frac{1}{4}$ or $\frac{3}{8}''$ have 18; pipes of $\frac{1}{2}$ or $\frac{3}{4}''$ have 14; pipes of from 1 to 2'' have $11\frac{1}{2}$; larger ones, 8.

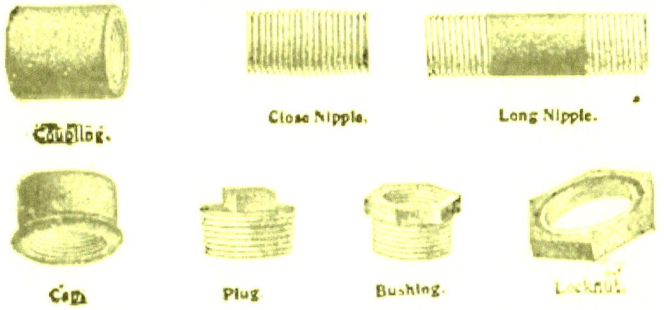

Coupling.   Close Nipple.   Long Nipple.

Cap.   Plug.   Bushing.   Locknut.

Q.—Name the principal features of the brickwork about a horizontal boiler?

A.—Binder bars, back stays, cleaning out doors, iron rollers and plates for the boiler lugs to rest on.

Q.—What is a globe valve?

A.—It is a valve in a round or globe chamber, used on boilers, engines, etc.

Q.—What are **thimbles** on boilers?

A.—They are heavy castings riveted on the upper shell of the boiler with planed flanges to which are bolted the safety valve and main steam pipe.

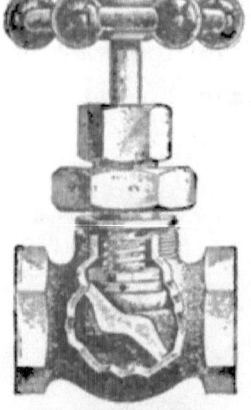

Globe Valve

Q.—Is a horizontal boiler placed level on its saddles?

A.—No, it is given a slight tilt (1½ inch) toward the back, so all the water can be drained out through the blow-off. This also insures having always water at the end opposite the gauge cock.

Q.—How are the sizes found of steam, water and gas pipes?

A.—By measuring their inside diameters.

Q.—How do you find the size of a boiler tube, flue or gauge glass?

A.—By the outside diameter.

Q.—In taking charge of a new plant, what is the first thing to do?

A.—Look after the water and steam pipes, also the valves connected with them.

Q.—Does water become lighter or heavier in a boiler under steam pressure?

A.—It becomes lighter per cubic foot as its temperature increases.

Q.—State as near as practicable the place to tap an **upright** boiler for the lower gauge cock?

A.—Two thirds the distance between the two flue sheets, measuring from the bottom flue sheet.

Q.—Where would you place the lower gauge cock in a **submerged tube** vertical boiler?

A.—From 2½ to 4 inches above the top flue sheet, according to the size of the boiler, so that the top ends of the tubes would always be submerged.

Q.—Where is the **water line** of a horizontal tubular boiler?

A.—From one and a half to two inches above the tubes.

Q.—Where is the **fire line**?

A.—On outside of shell and in line with the upper row of tubes.

Q.—Where is the lower gauge cock in a horizontal tubular boiler?

A.—An inch and a half to two inches above the upper row of flues.

Q.—Where is the water pipe tapped in a boiler head for a **water-combination column?** Where is the steam pipe tapped, and what size of pipe is used for making the two connections?

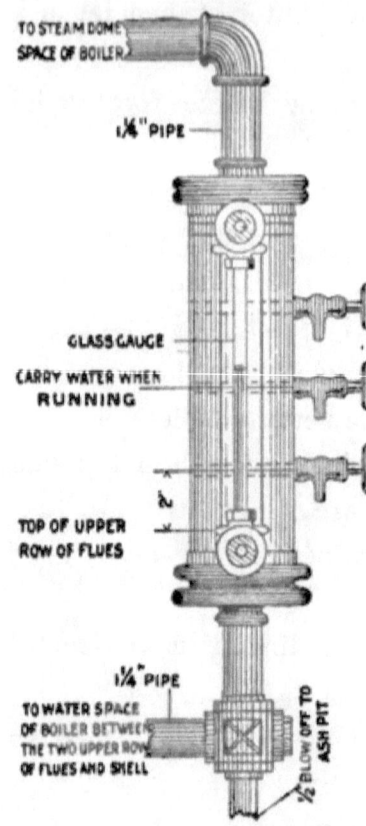

A.—The water pipe is generally tapped centrally between the two upper rows of flues and the shell of boiler. The steam pipe is tapped in the top of the shell or in the dome. The connecting pipes should not be smaller than 1¼ inch diameter.

Q.—How often would you blow out the gauge glass during the day?

A.—About four times, or as often as necessary.

Q.—Is a glass gauge always perfectly reliable?

A.—No. The gauge cocks must be tried even if a glass gauge is used.

Q.—Where is the **safety plug** usually placed in a water tube or flue boiler?

A.—In water tube boilers they are generally placed four inches above the bottom of the drum, and not in the tubes. In flue boilers it is sometimes screwed in the top of one of the upper flues, but of late it is the custom to tap the crown of the shell about 15 inches back of the dome and there insert a half-inch pipe, reaching to within ¾ of an inch of the flue line. The top of this pipe is tapped into a brass chamber, in the

top of which the safety fuse plug is screwed in.

Q.—How does this arrangement work?

A.—When the water in the boiler falls below the lower end of the safety plug pipe (indicated by a dotted line in the cut), the dry steam enters it, passes into the chamber and fuses the plug, the steam escapes and gives warning. See also page 68.

Q.—At what temperature does the plug fuse, and what is it made of?

A.—It is made of Banca tin which fuses at 420° F.

Q.—What causes **channeling** and **grooving** in a boiler?

A.—They are caused by the mechanical action produced by unequal expansions and contractions.

Q.—Where would you feed water into a boiler to prevent grooving, etc.?

A.—Feed near the water level of the boiler instead of near the bottom.

Q.—Which is the best **arrangement of the feed pipe?**

A.—It should enter the front head just above the tubes and a few inches away from the shell. It should then extend back to within a foot or so of the back head, then cross over and discharge on the opposite side, downward, between the tubes and shell. In this way the feed water becomes well heated before discharging into the boiler.

Q.—How large should a feed pipe be?

A.—According to the size of the boiler, from 1 to 1½ inches.

Q.—How large should the **blow-off pipe** be?

A.—Ordinarily 2 to 2½ inches diameter.

Q.—Where should the blow-off pipe be attached to the boiler?

A.—Underneath its back end. The shell should be re-enforced with a flange riveted on, and the pipe should be protected from the action of the flames and hot gases from the furnace by a fire brick stand.

Q.—Why is malleable iron used for the elbows in the fire?

A.—Cast iron ones would burn or break.

Q.—Is it dangerous to empty a boiler when the tubes or flues are hot?

A.—Yes; and it is also dangerous to hastily fire up a boiler, because where the draught and combustion are sufficient for a white heat, the plates, no matter how good they may be, cannot with certainty resist the terrible heat.

Q.—State causes of **defective circulation**.

A.—It is caused by flues being too close together, scale thickening on them, and flues set zigzag.

Q.—What construction of a boiler would be considered successful and economical?

A.—For a tubular boiler place the flues in vertical rows, leaving out the center row; good circulation is when water goes down in the center and rises at the sides where the heat strikes it.

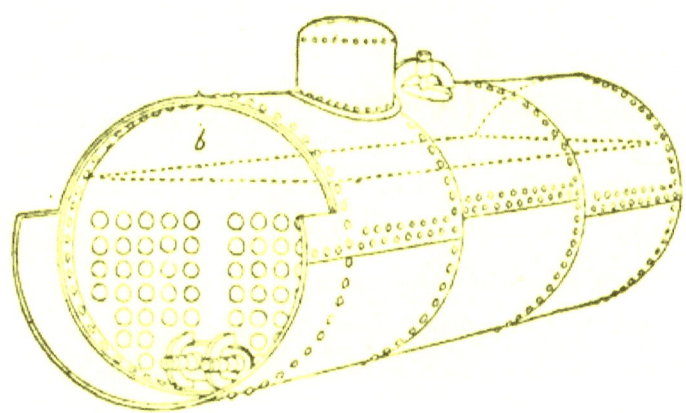

Q.—How much steam-space is there in a boiler?

A.—About ¼ of the internal capacity. (In the cut the water surface is indicated by dotted lines, and the height of the steam space by b.)

Q.—Give the space between the flues of a well-made boiler?

A.—The proper space should be half the diameter of the flue. (See the cut above.)

Q.—What amount of water in weight can be evaporated by one pound of good coal?

A.—The average is from six to ten pounds.

Q.—What waste of heat is there if 1-16 inch of scale is in the boiler?

A.—Some of the best authorities claim from 10 to 15 per cent of fuel, and in this proportion upward according to thickness of scale.

BOILER EXPLOSIONS

Q.—What causes a boiler to explode?

A.—It may be one or several of various causes. Defects in material or construction, or improper management account for most explosions.

Q.—What is the scientific explanation of an explosion?

A.—A boiler explodes when the pressure within exceeds its resisting power.

Q.—What decides a boiler's resistance?

A.—The strength of its weakest spot. It is there that an excessive pressure breaks through first.

Q.—Why are the parts surrounding the weakest spot affected?

A.—The break decreases their resisting power, while the shock and the sudden increase in the generation of steam manifold the pressure.

Q.—Does all the water instantly change to steam?

A.—No, but with a speed increasing at such a rapid rate that it seems instantaneous.

Q.—Is **low water** often a cause of explosion?

A.—Yes, when the engineer tries to fill the boiler quickly, instead of very slowly. If a large amount of cold water suddenly enters a hot boiler with a high pressure, too much of it changes to steam, at once raising the pressure beyond the resistance of the boiler.

Q.—Is it proper, then, to feed water into a boiler when the water is out of sight?

A.—*Under no circumstances.*

Q.—What would you do?

A.—I should immediately draw the fire, if a light one; if a heavy one, I should cover it over with wet ashes to deaden the heat.

Q.—Why not draw out a heavy fire?

A.—Because it would make more heat by raking.

Q.—What would you do if the water was too high in the boiler?

A.—Carefully open the blow-off and let out one gauge of water.

Q.—What injury and danger are caused by heavy scale in a boiler?

A.—The heat from the boiler plates is not communicated to the water directly, but through the incrustation, a bad conductor. This necessitates an overheating of the plates, which deteriorates and weakens them rapidly.

Q.—What is a bagged or blistered boiler?

A.—A bag is a bulging out of the plate; a blister is a bulging out not of the whole plate, but of the outer layer split from the inner. These defects are caused by too much sediment or scale. They weaken the boiler very much.

Q.—How are these defects remedied?

A.—By cutting the bagged or blistered piece out, and riveting a hard patch on the inside of the boiler.

Q.—Why on the inside?

A.—Because if put on the outside, the hole would form a pocket for sediment.

Q.—How would you find the safe working pressure of a boiler?

A.—Multiply twice the thickness of shell by the T. S. stamped on boiler plate and divide answer by 6 times diameter of shell in inches. If double riveted multiply by .70; if single, by .56. (Ans. in lbs. of pressure gives the safe load, at which the safety valve is set.)

Q.—How is the safe working pressure found in cylindrical boilers?

A.—Multiply one-sixth of the lowest tensile strength by the shell's thickness (expressed in parts of an inch) at the thinnest part, and divide the product by the inside radius (half diameter) in inches. The answer will be the pressure allowable per square inch of surface for single riveted; if **double riveted add 20 per cent.**

Q.—Do these two rules give the same result?

A.—No, but they are both used by different engineers, and are both claimed to be serviceable.

Q.—Which do you consider the safer, drilled or punched holes in boilers, for rivets, etc?

A.—Drilled holes.

Q.—Describe a good way of keeping a boiler clean?

A.—Every boiler should be supplied with a surface blow-off, as a large percentage of the foreign matter held in suspension in water rises at the boiling point and can then be blown off before it has had time to deposit on the surface and flues. If not blown off, the heavier particles will be attached to each other until they become sufficiently heavy to fall to the bottom, when they will be deposited in the form of scale, covering the whole internal surface of the boiler below the water line.

Q.—Where is the surface blow-off tapped?

A.—It is tapped in the crown of the boiler and its pipe is bent so as to lie even with the average water level. When the valve is opened, the outrushing steam carries the surface water and any light matter floating on it along into the catch basin. This device is usually called the skimmer.

Q.—When is the proper time and how would

you blow out a boiler for cleaning purposes?

A.—Allow the furnace and boiler to cool down, open blow-off so the water and mud will escape, then wash out with a hose. Scrape the flues if possible, pull out all the sediment and scale left on the bottom of the shell with a long-handled hoe through the hand hole of the boiler and thoroughly rinse with water.

Straightway Blow-off Valve

Q.—Suppose a boiler was found badly corroded and pitted internally along the water line, and covered with a heavy deposit of sediment, baked on hard, what should be done?

A.—Get inside the boiler and thoroughly scrape the shell, getting down to the sound plate, then with a stiff wire brush thoroughly oil or paint the corroded portion with red lead and boiled linseed oil, three coats.

Q.—What are the causes of "foaming"?

A.—Foaming comes from various causes, such as the mixing of water with steam, high water, irregular firing or feeding, impure or greasy water, too small steam space, dirty boiler, changing of water, etc.

Q.—How is it known when a boiler foams?

A.—It can be seen in the gauge glass by the water suddenly moving up and down, or by the sputtering at the gauge cock.

Q.—Can foaming be overcome?

A.—Yes, by partly closing large valves, opening fire doors and feeding water into boiler.

Q.—What is meant by a boiler's "priming"?

A.—Entrance of water together with steam into the steam pipe, caused by high water, narrow steam pipe and sudden opening of valve.

Q.—How would you remedy it?

A.—By opening the valve slowly, or lowering water in the boiler.

Q.—How would you **gasket** a steam joint so the gasket can always be taken out and replaced without injuring it?

A.—By rubbing a little graphite and oil between the face of the flanges and the gasket, both sides.

Q.—How would you remove a man or hand hole plate from a boiler?

A.—Simply loosen nut, remove brace (dog or crab), and turn the plate to narrow side and take out.

Q.—Why is a hand or man hole plate made oval instead of a true circle?

A.—So they can be taken out and put in and new gaskets put on.

Q.—When you have a battery of two **boilers** or more and one boiler has 80 lbs. steam pressure and the rest are cold, how would you proceed to connect them together?

A.—Simply fire up the cold boiler and raise steam pressure to equal the one to which you wish to connect it. Never under any circumstances turn high into low pressure or hot into cold, because the sudden expansion may cause a serious rupture and may cost you your life.

Q.—What is the first thing you would do on entering the boiler room in the morning?

A.—See how much water is in the boiler by trying the gauge cocks, etc.

Q.—Then what would you do?

A.—Start a fire if I had one or two gauges of water.

STEAM HEATING

Q.—How would you open a steam valve to supply steam to a building for heating in the morning?

A.—Open the valve slightly and wait until the pipe stops pounding, then gradually open a little more; it saves joints, gaskets, pipes, etc. When opening valves, make sure that the valve at the end of the return pipe is open until hot, then close it.

Q.—How is the amount of pipe required for properly heating a room calculated?

A.—By the following rules:

One cub. foot of boiler to every 1,500 cub. feet of space. One H. P. of boiler to 40,000 cub. feet of space. One superficial foot of steam pipe to six superficial feet of glass in windows. One superficial foot of steam pipe to 100 sq. feet of wall, ceiling or roof. One sq. foot of steam pipe to 80 cub. feet of space.

Q.—How do you find the heating surface of a radiator?

A.—Multiply the total length of all the pipes by the outside circumference in inches, and divide by 144. The answers give the square feet.

Q.—Which is the best way to thaw out frozen steam pipes?

A.—By laying some old cloth or waste on the pipe and pouring on boiling water, the pipe can be thawed out in 10 minutes.

SMOKE AND CHIMNEY

Q.—Would it be proper to have the chimney rough inside?

A.—No; it should be as smooth as can possibly be made, and the area a little larger toward the top than at the bottom (inside).

Q.—How much larger should the space be where the smoke or gases return through the flues than the grate surface?

A.—It should be one-fifth larger in area than the grate surface.

Q.—Where would you consider the proper place to close in against the sides of an externally fired boiler with brick (fire line)?

A.—About in line with the center of upper row of flues all along the full length outside of boiler.

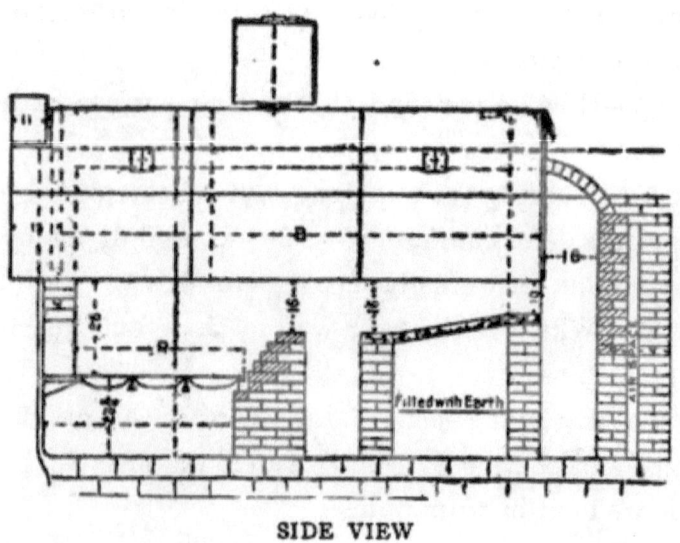

SIDE VIEW

This cut shows the proper way of enclosing a boiler in brickwork. The figures give the distances in inches.

Q.—Where does the greatest effect of the fire on the bottom of an externally fired horizontal boiler take place?

A.—Just back of the bridge wall.

Q.—From where is the height of a chimney measured?

A.—From the top of the grate.

Q.—What makes a chimney draw?

A.—The difference between the weight of the column of heated gases within and an equal column of cooler air without.

Q.—Upon what does the draught capacity of a chimney depend?

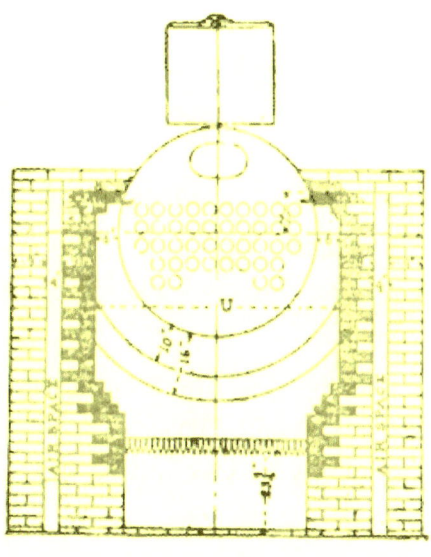

END VIEW

A.—Upon its height, cross section area, and upon the temperature.

Q.—State the size of chimney necessary to fully relieve the tubes or flues of a boiler or boilers of smoke, and give height?

A.—The chimney should be one-fifth larger in area than all the tubes or flues combined, so as to afford an ample passage for all the gases. The top should project at least 10 feet above the highest building in the immediate vicinity, to avoid all downward currents of the atmosphere.

BOILER TESTING

Q.—Is the hydraulic test or the hammer test better? and why?

A.—The hammer test is always reliable because a flawless metal gives a clear sound, and every part, inside and out, is examined by itself. In the hydraulic test a boiler may get strained, and when heated afterward, the expansion may bring out a leak. Government and insurance inspectors employ the hammer test.

Q.—How do you find a broken or loose stay or rivet?

A.—By holding a hammer against one side and striking the other side with another hammer. Any looseness can be discovered in this way by the feeling.

Q.—If tested by the hydraulic test how much pressure is sufficient to test the boiler so as to carry a certain amount of steam pressure?

A.—The hydraulic pressure test should be one-half more than the steam pressure to be carried, viz.: If steam pressure is to be 80 lbs. the hydraulic test should be 120 lbs.

Q.—What is it that **ruptures** a boiler?

A.—The pressure within it, and strains caused by unequal expansion and contraction.

To avoid this trouble it is necessary to

exercise great care in raising steam. The fire should be increased gradually and the boiler have at least four inches of water above the top row of flues so the temperature may be gradually raised.

Q.—Is it injurious to a boiler to open the fire doors often and suddenly cool the fire and sheets?

A.—Yes; it is very unsafe.

FEED REGULATION

Q.—Suppose you had a battery of three boilers and the only valves near the boilers on the feed pipe were check valves, how would you feed the boilers evenly without using globe valves between the checks and boilers?

A.—First, fire the boilers evenly; second, keep the pumps running steady, and if one boiler should happen to receive more water than the others use the blow-off valve of that particular boiler and regulate the height by it.

Q.—Can uneven feeding be prevented?

A.—Yes, by partially closing the stop valves of the boiler or boilers with high water, and, if necessary, by opening the stop valves of the low water boilers a little more.

BOILER HORSE POWER

Q.—How can you find the amount of water evaporated in a boiler?

A.—Take the mean between the widths of the

two levels at the beginning and at the end of a space of 15 minutes, as indicated by the glass gauge. (See cut, pp. 42, 45.) Multiply the constant length of water surface with this mean width and multiply their product by 4. This gives the amount evaporated in cubic inches. Dividing the result by 1728, you get the answer in cubic feet. This test is not recommended, though used.

Example: If the glass gauge shows a difference of one inch, we measure across the face of the boiler half an inch above the last level. If this measures 48 inches and the boiler is 14 feet long (=168 in.), we have $48 \times 168 = 8064$. Multiplied by $4 = 32,256$ cub. in. per hour, or $18\frac{2}{3}$ cub. feet.

Q.—Can you know from the amount of water evaporated in one hour, how many horse-powers have been developed?

A.—Yes.

Q.—How many cubic feet of water evaporated in one hour equals a horse-power?

A.—One-half cubic foot, $3\frac{3}{4}$ gallons or 864 cubic inches.

Q.—Then, in our example, how many horse-powers were indicated?

A.—Two times $18\frac{2}{3}$, $=37\frac{1}{3}$ H. P.

Q.—How can you find the horse-power of a tubular boiler by the heating surface?

A.—First find the number of square inches of heating surface around boiler shell from fire line to fire line and in the flues; divide by 144 to get square feet; divide quotient by 15 if horizontal tubular, and by 12 if locomotive or vertical boiler to get H. P.

FEED-WATER HEATER

Q.—How many types of feed-water heaters are there?

A.—Two. The open heater and the closed.

Q.—What is the difference between them?

A.—In an open heater the exhaust steam comes directly in contact with the feed-water, in a closed heater it does not.

Q.—What is the object of a feed-water heater?

A.—To save fuel by making use of the exhaust steam from the engine to heat the feed-water.

Q.—At what temperature will a heater deliver water to a boiler?

A.—That depends upon the type of heater and other conditions. A good heater of ample proportions should raise the temperature of the feed-water up to 200° F., or higher.

Q.—What else is a heater good for besides heating the feed-water?

A.—It also purifies the water by extracting the scale-producing matter, and also the mud.

Q.—In using an open heater is there any danger of flooding the cylinder, and if so, how?

A.—If there should be any stoppage of the outflow of feed-water, it would flood the cylinder through the engine exhaust pipe, and perhaps cause a wreck.

TENSILE STRENGTH

The tensile strength of metals is the load that would break a bar of one inch area in cross section if applied in the direction of its length.

For a test of the tensile strength of iron or steel boiler plates, narrow strips are sheared from plates selected at random from a pile of them rolled at the same time—we will say the plates are steel and a quarter of an inch in thickness. These strips are at the middle reduced to a quarter of an inch each way (square).

Suppose the testing machine pulls the first of 4 strips asunder at 3,999 lbs. register, the second breaks at 4,001 lbs. and the last two at exactly 4,000 lbs. each. Adding these all together we have the sum of 16,000 lbs., which, divided by 4, the number of strips tested, gives us 4,000 lbs. as the mean breaking strain of a quarter square inch of sectional area of steel.

Multiply this by the number of quarter square

inches in 1 square in. and we have the tensile strength in 1 square in. of section. There being 16 quarter in. square in 1 square in. would give 16 times 4,000, which equals 64,000 lbs. for a bar having 1 square in. of sectional area, which would be about the average tensile strength of first quality steel.

After tensile strength is found, all the plates are stamped T. S. in that particular batch, and underneath stamp 64,000. Sheets not stamped should not be rated at more than 48,000 lbs. T. S.

STEEL AND IRON

Q.—What is steel?

A.—Steel is a variety of iron containing from one-half of one per cent to one and a half of one per cent of carbon.

Q.—What is iron?

A—Iron is a metal, the most abundant and the most important of all. It contains always impurities, such as magnesia, sulphur and phosphorus. It is hardly anywhere found native, but must be manufactured from ore. *Cast iron* is brittle and hard. *Wrought iron*, obtained by *puddling*, is softer and malleable.

Q.—What are the principal advantages of steel over iron?

A.—Greater elasticity and hardness, which by tempering may be increased to any desired degree.

POP AND LEVER SAFETY VALVES

Q.—Of what use are safety valves?

A.—They are to release the boiler automatically of all steam pressure above a certain point.

Q.—Are there more than one kind of safety valves?

A.—Yes—the old lever and the spiral spring safety valve.

Q.—When steam is heard issuing from a safety valve, does it signify danger?

A.—No; it is a signal of safety. It shows the valve is in working order and, if properly set and adjusted, it is a sure protection against trouble

Q.—How do you set a pop safety valve?

A.—In setting the valve shown on page 63, remove the cap H., and turn the set bolt O up or down, to decrease or increase the pressure.

Q.—How is the amount of reduction regulated?

A.—Remove the set screw D (Fig. 1) from the lower part of the case M, insert a pointed instrument in the screw hole, and with it turn down (to the left) the set ring, increasing the amount of loss, or up (to the right) for decreasing the amount. Then replace the set screw which holds the ring in position.

V is the valve nut into which O is screwed. B is the valve. N is the upper cap over spring casing K inside casing M. S is the upper spring cap, R the lower. T is the testing lever, C the main casing, E the bolt bushing, F the bushing jam nut, A the guide for valve disc, J the guide for lower valve stem.

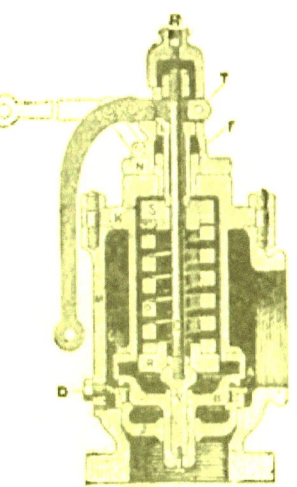

Fig. 1

Q.—How large a loss is it usual to have?

A.—Three or four pounds. A valve can be set to lose less than half a pound in popping.

Q.—What new device is there for deadening the sound of the pop valve?

A.—The muffler attachment. (Fig. 2.)

Q.—How is the Muffler Valve adjusted?

A.—It can be adjusted on top without removing from the dome. In order to adjust either the pressure or the blow-down, first remove the muffler I; this exposes the compression screw G, adjustable nut M, crosshead L, locking latch O, and check nut H. By loosening the check nut H and screwing down the compression screw G, you increase the pressure, and the reverse for lessening the pressure. (As a general rule from 1-16 to ¼ turn will change the pressure of valve five lbs.

either way.) By raising the locking latch O and screwing down on the adjusting nut M one notch you will reduce the blow-down one pound, and the reverse increases it one pound.

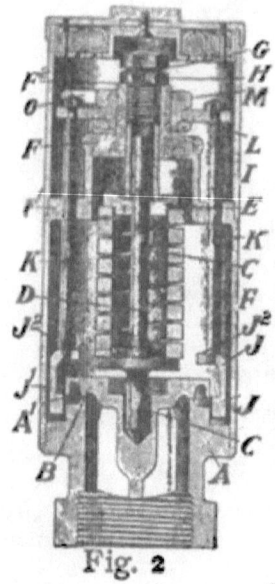

Fig. 2

A base, A^1 valve seat, B valve, C spindle, D spring, E follower, F F^1 main casting, F^2 thread hub, G compression screw, H check nut, I muffler, J the regulating ring, J J^1 lugs on ring, K parallel rods, L cross-head, M adjusting nut, O locking latch.

Q.—Give proper size of safety valve for a boiler having 25 sq. feet of grate surface, allowing for 70 lbs. pressure?

A.—For each foot of grate surface 22.5 feet boiler heating surface is allowed; $25 \times 22.5 = 562.5$. For the water in the boiler we allow 8.33 (the weight in lbs. of one gallon), which, added to the given pressure, gives 78.33. 562.5 divided by 78.33 equals 7.18 sq. inches area, or a 3-inch diameter.

Q.—How would you figure the pressure under a 3-inch safety valve with 75 lbs. boiler pressure?

A.—Three times 3 equals 9 inches, times .7854 equals 7.068 area, times 75 equals 530.1 lbs.

Q.—What is the United States government rule about the relative areas of grate and safety valve?

A.—One square inch area of lever safety valve to 2 square feet of grate surface.

Q.—State the general allowance among inspectors?

A.—One inch area of safety pop valve to 3 square feet of grate surface.

Q.—Find area of a pop valve 3½ inches in diameter? (See table of areas, page 235.)

A.—Multiply diameter by itself and then by decimal .7854; answer is the area, less decimals.

Q.—When calculating the load on a safety valve, is allowance made for the atmospheric pressure on top of valve?

A.—No; because it is present everywhere, inside and outside the boiler, and may be left out of the calculation entirely.

Q.—Are the spring safety pops calculated when set?

A.—No; as a rule they are set by a test steam gauge.

Q.—What is done, if in such a test the needle does not show true at 100 lbs. pressure?

A.—It is pulled off the pin, and then put back in the right position.

Q.—What is meant by a *strong* or *light* needle?

A.—It is termed strong when at the test it shows less than the true pressure, and it is called light when it shows more.

Q.—What is it that keeps the face of the

valve and the seat in line (opposite) and causes the rise and fall to be even and true?

A.—The valve spindle.

Q.—What is the point of contact?

A.—Where the valve and its seat meet.

Q.—At what angle is the edge of the valve and its seat beveled?

A.—At an angle of 45 degrees.

Q.—How is it known when the safety valve is in good working order?

A.—By the steam and gauge. Let the steam pressure rise enough to just move the safety—no more—and note the correspondence between the gauge and safety valve.

Q.—Is there another way?

A.—Yes—raising the valve by hand.

Q.—How do you find the exact place where to place the ball (weight) on the long lever of the safety valve?

A.—By applying the laws of leverage. (Page 241.)

Q.—How do they apply in a safety valve?

A.—The bar holding down the valve is a lever of the third kind, the pivot representing the fulcrum, the valve representing the power, and the ball representing the weight. (See cut, page 68.)

Q.—Give the rule for calculating the distance from the fulcrum at which a given weight must be set to cause the valve to blow at any specified pressure.

SAFETY VALVE

A.—1. Multiply the area of the valve in square inches by the pressure in pounds per square inch. Call this product "number 1."

2. Multiply the weight of the lever in pounds by the distance in inches of its center of gravity from the fulcrum; divide the product by the distance in inches from the center of the valve to the fulcrum; add to the quotient the weight of the valve and spindle. Call this sum "number 2."

3. Divide the distance in inches from the center of valve to fulcrum by the weight of the ball in pounds, and call the quotient "number 3."

4. Subtract "number 2" from "number 1," and multiply the difference by "number 3"; the product is the answer.

Example: Given: diameter of valve 4 inches; distance from fulcrum to center of valve 4 inches; weight of lever 7 lbs.; distance from fulcrum to center of gravity of lever 15½ inches; weight of valve 3 lbs.; weight of ball 108.24 lbs. Blowing-off pressure 75 lbs.

Area of 4″ valves = 12.566 square inches

$$75 \times 12.566 = 942.45$$
$$\frac{7 \times 15.5}{4} + 3 = 30.125$$
$$4 \div 108.24 = .0369$$
$$942.45 - 30.125 = 912.325$$
$$912.325 \times .0369 = 35.66 \text{ inches. Ans.}$$

Q.—How do you find the pressure at which a safety valve will blow off when the weight and its position are known?

Iron Body Cross Safety Valve, Flange Ends.

A. — Divide the fulcrum into the length of lever, multiply by weight of ball, add weight of lever, valve and stem and divide by area of valve.

Q.—How is the total weight of lever, valve and stem found?

A.—The easiest way is to tie the stem with a string to the lever and attach a spring (scale) balance to the lever and valve, directly over the center of the valve, and weigh them in place.

AUTOMATIC EXTINCTION OF FIRE BY STEAM AT LOW WATER

In this device, recently patented in Vienna, a pipe reaches through the top of the boiler down to low water mark, so that steam will enter it as soon as the water falls below the mark. In the upper end of the pipe a safety fuse is melted by the steam, opening connection with a second pipe, which leads into the fire-box, where the steam extinguishes the fire.

The fuse being a ring between a conical valve and its seat, the valve can be screwed down on the valve seat, as soon as the fuse is melted out, and a new fuse put in at any convenient time.

A whistle or bell is easily connected with the apparatus to give alarm. The air in the pipe first mentioned is exhausted through a stop-cock, after the boiler is heated.

INJECTORS

Q.—What is an injector and its use?

A.—It is a substitute for a pump and is used in feeding a boiler with water.

Q.—How is it that an injector forces water into a boiler against the pressure of the steam operating it?

A.—The water and steam mingling at the combining tube, the steam jet is condensed, converted into a water jet. This water jet has a much smaller cross section area than the steam jet had, and as the energy of the steam jet is retained entire, a greatly increased velocity results.

Q.—What forces the boiler check valve open?

A.—The pressure of the water in the delivery pipe.

Q.—State the velocity of steam passing through an inch pipe at 100 lbs. pressure?

A.—Two thousand feet per second.

Q.—Where would you look for trouble if the injector stream broke and the same injector always before worked well?

Monitor Injector

NAMES OF PARTS

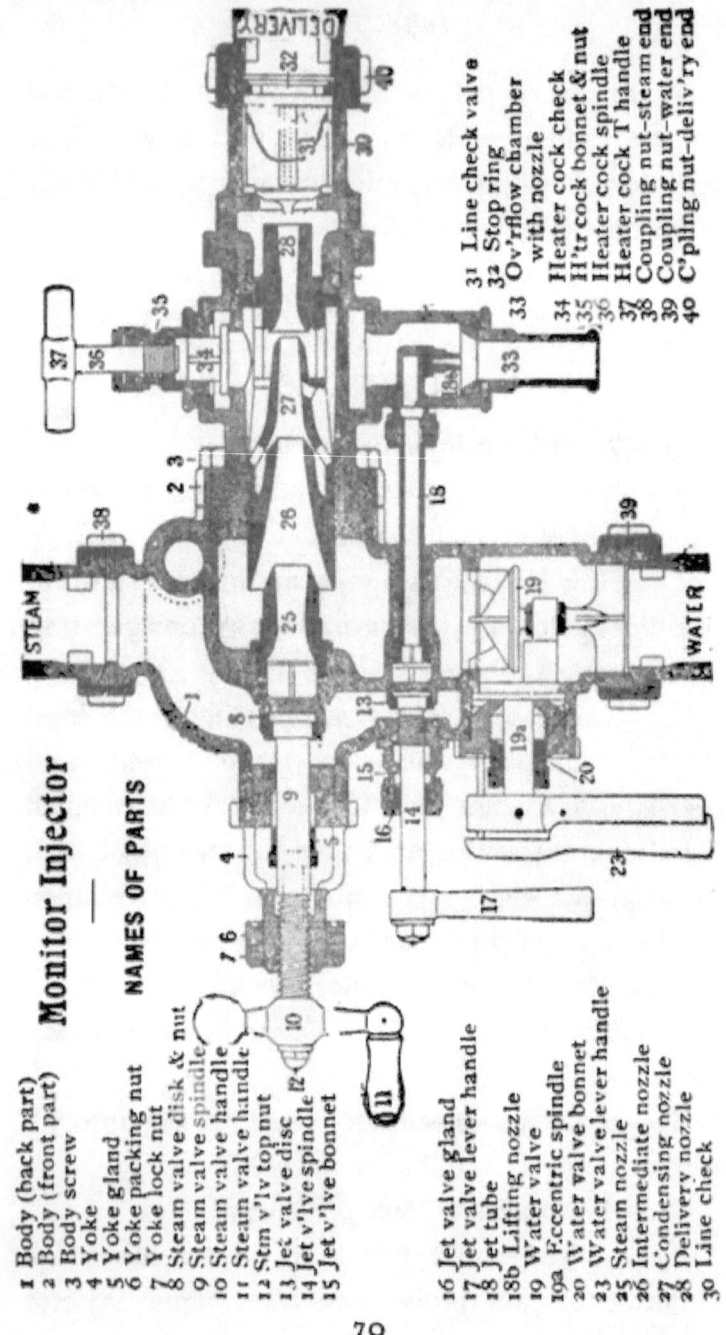

1 Body (back part)
2 Body (front part)
3 Body screw
4 Yoke
5 Yoke gland
6 Yoke packing nut
7 Yoke lock nut
8 Steam valve disk & nut
9 Steam valve spindle
10 Steam valve handle
11 Steam valve handle
12 Stm v"lv topnut
13 Jet valve disc
14 Jet v"lve spindle
15 Jet v"lve bonnet
16 Jet valve gland
17 Jet valve lever handle
18 Jet tube
18b Lifting nozzle
19 Water valve
19a Eccentric spindle
20 Water valve bonnet
23 Water valve lever handle
25 Steam nozzle
26 Intermediate nozzle
27 Condensing nozzle
28 Delivery nozzle
30 Line check
31 Line check valve
32 Stop ring
33 Ov'rflow chamber with nozzle
34 Heater cock check
35 H'tr cock bonnet & nut
36 Heater cock spindle
37 Heater cock T handle
38 Coupling nut-steam end
39 Coupling nut-water end
40 C'pling nut-deliv'ry end

A.—At the water and steam supply.

Q.—Of what use is a steam nozzle?

A.—It is for the actuating steam jets to pass.

Q.—Where is the combining tube?

A.—In the casing of the injector where the steam and water mix.

Q.—Where is the delivery tube of an injector and what is its use?

A.—It is where the maximum velocity of the stream is attained, and the jet overcomes the back pressure from the boiler.

Q.—Are injectors divided into classes; if so, state how many?

A.—Yes; they are divided into two general classes, the lifting and non-lifting.

Q.—Can these two classes be subdivided?

A.—Yes; they may be divided into six, namely, single tube, double tube, self-adjusting, restarting, open or closed overflow injectors.

Q.—Is it a good idea to turn on more steam after overflow has been shut off?

A.—No; it will cause the injector to break the stream.

Q.—State some of the principal causes that make an injector's stream break?

A.—Not enough water supply, straws, chips, mud, cinders, leaky joints, overheated water, bad strainers, corrosion in the injector casing and low steam pressure.

Q.—What rules are used for determining the proper size of an injector for different boilers?

Rule 1. For Vertical Tubular Boilers—reduce all dimensions to inches and multiply the circumference of the fire box by its height above the grate; multiply the combined circumference of all the tubes by their length; next subtract from the area of the lower tube sheet, the area of all the tubes and add the remainder to the sum of the area of the tubes and shell and divide total by 144, and the quotient will be the number of square feet of heating surface.

Rule 2. For Horizontal Tubular Boilers—reduce all dimensions to inches and multiply two-thirds of the circumference of the shell by its length; multiply the length of the tubes by their combined circumference; next subtract from two-thirds of the area of both heads the combined area of the tubes and add the remainder to the sum of the tubes and shell, divide total by 144, etc.

Rule 3. For Water Tube Boilers.—Proceed to find the area of all heating surfaces exposed to the radiation of gases from the furnace of boiler, and if area is in inches, divide by 144, etc., as above.

After finding the heating surface, as per rule 1, 2 or 3, divide by 30 (see page 150) to get the horse power, and allow 10 gallons of water per hour for each horse power.

Q.—What would be a short rule then?

A.—If H. P. is known, multiply number by 10 to find number of gals. of water the injector should deliver per hour. If H. P. is not known, multiply number of sq. feet of heating surface by 3.

FEED PUMPS

Q.—Name the different kinds of pumps used daily for boiler feeding, etc.?

A.—Single action, with two valves, receiving and discharging; the double action, with two or more discharging valves. The latter receives and discharges water at both ends of water cylinder and has a steam cylinder attached to work the pump. The duplex is a combination of two double action pumps all cast together side by side.

Q.—What are the relative proportions of steam and water cylinders of feed pumps?

A.— The steam cylinder is 1-3 larger in diameter than the water cylinder.

Q.—In setting up a steam pump, how is it leveled?

A.—By leveling the discharge valve seat lengthwise and crosswise.

Q.—Give rule to find area of a steam piston in connection with a pump?

A.—Multiply the area of water plunger by 2.

Q.—Of what are pump valves made?

A.—Hard or soft rubber, brass and sometimes vulcanized fiber and wood.

Q.—Can you give a short rule to find the **sizes of steam pipes for cylinders?**

A.—Divide the area of steam piston by 64 for steam pipe, and by 32 for exhaust. Divide the area of plunger by 3 for discharge pipe, and by 2 for suction pipe.

Q.—Suppose you had a duplex pump, size 8 in. water by 10 in. steam by 12 in. stroke and 3 in. diameter plunger rod, making 100 ft. piston travel per minute, **how many gallons of water** would the pump deliver, having full supply of water?

A.—First find the area of plunger face—8 times 8 in. equals 64, multiplied by .7854 equals 50.2656, by 12 in. stroke equals 603.1872, by 4 cylinder ends equals 2412.7488 cubic in. Now subtract the cubic contents of 3 in. diameter plunger rod 12 in. stroke in one end of each water cylinder from the total cubic inches and divide by 231, which gives gallons for one stroke, 4 ends. This multiplied by 100 piston travel gives total. Three multiplied by 3 equals 9, by .7854 equals 7.0686, by 12 equals 85.032, by 2 equals 170.064, subtract from 2412.7488 equals 2242.6840, divided by 231 cubic inches in a gallon equals 9.708 gals. one stroke, multiplied by 100 equals 970.8 gals. per minute. This rule holds good on other pumps.

Q.—Give quick rule to find quantity of water

pumped in one minute, pump making 100 ft. of piston speed per minute?

A.—Multiply the diameter of the water plunger by itself, then multiply the product by 4. Answer gives gallons for one pump; if for two pumps multiply answer by 2 and so on.

Q.—How could the **horse-power** be found necessary to pump water to a given height?

A.—Multiply the total weight of water in pounds by the height in feet and divide by 16,500. This allows for water friction and steam loss.

Q.—How are the **steam valves of duplex pumps** set and adjusted?

A.—Remove the valve chest cover, place the rocker arm plumb (reach arm), then see how the valve on opposite cylinder is for lead; if equal at both ends, the valve is set, if not, adjust the jamb nuts to suit. Do the same on the other pump.

Q.—Does the duplex pump exhaust its steam through the same port that it enters and thence through the cavity under the valve to the exhaust chamber as in the common slide-valve engine?

A.—No. It has a separate steam and exhaust port at each end. (See cut, page 76.)

Q.—How does it work?

A.—The piston covers and closes the exhaust before it reaches the end of its stroke, and the steam left in the cylinder acts as a cushion. At the end of stroke the steam valve opens and the

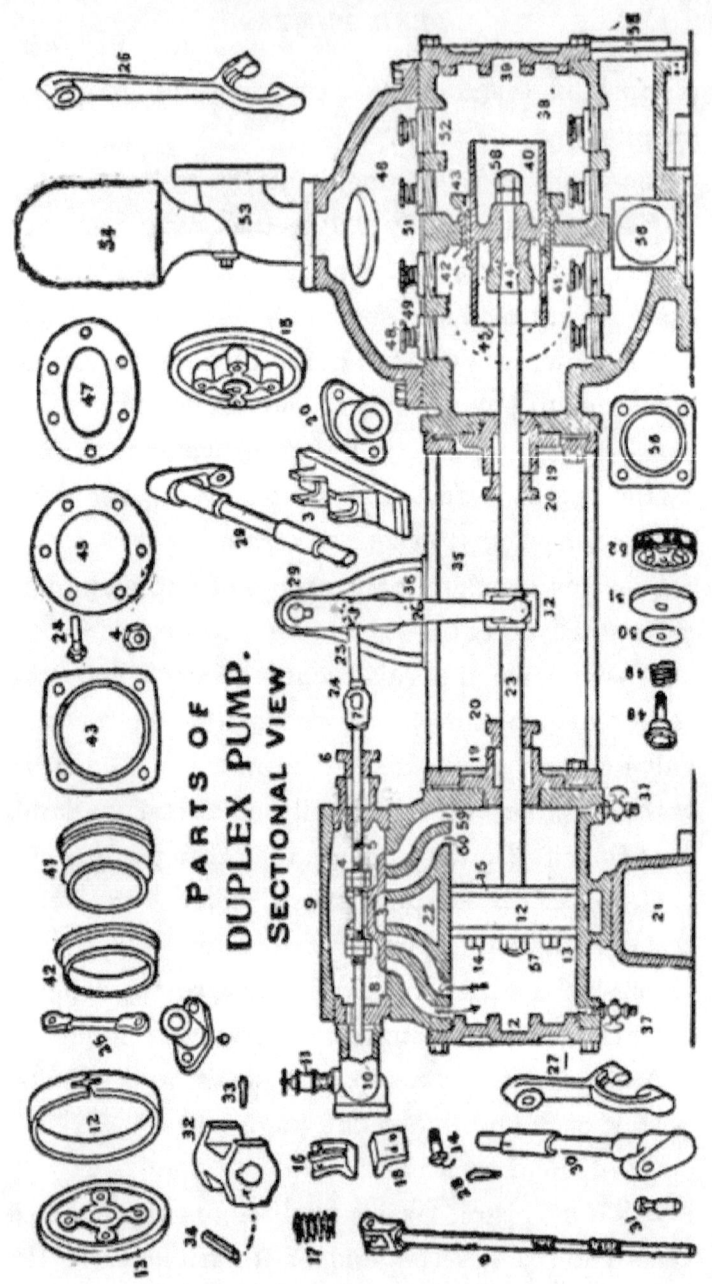

FEED PUMPS

Numbered List of Parts Illustrated on the Opposite Page.

1. Steam Cylinder.
2. Cylinder Head.
3. Slide Valve.
4. Valve Rod Nut.
5. Valve Rod.
6. Valve Rod Gland.
7. Valve Rod Head.
8. Steam Chest.
9. Steam Chest Cover.
10. Steam Pipe.
11. Lubricator.
12. Piston Ring.
13. Piston Follower.
14. Follower Bolts.
15. Piston Body.
16. Piston Tongue.
17. " Tongue Spring.
18. " Tongue Bracket.
19. " Rod Stuffing Box.
20. " Stuffing Box Gland
21. Steam Cylinder Foot.
22. Exhaust Outlet.
23. Piston Rod.
24. Valve Rod Head Pin.
25. Rod Link (long & short)
26. Long Lever.
27. Short Lever.
28. Rock Shaft Key.
29. Upper Rock Shaft.
30. Lower Rock Shaft.
31. Crank Pin.
32. Spool.
33. Spool Position Pin.
34. Spool Key.
35. Cradle.
36. Cross Stand.
37. Blow Cock.
38. Water Cylinder.
39. Water Cylinder Head.
40. Plunger.
41. Plunger Ring.
42. Casing.
43. Binder.
44. Plunger Hub.
45. Water Cyl. Hand Plate
46. Force Chamber.
47. "   " II'nd Plate
48. Valve Guard.
49. Valve Spring.
50. Brass Valve Plate.
51. Valve.
52. Valve Seat.
53. Delivery Tee.
54. Air Chamber.
55. Suction Screw Flange.
56. Suction Hand Plate.
57. Piston Nut.
58. Plunger Nut.
59. Steam Ports.
60. Exhaust Ports.

live steam forces the piston back. The same at both ends.

Q.—Why does the piston with the long lever move in the same direction as the opposite valve, while the other piston with short lever and the opposite valve rod move in opposite directions to each other?

A.—The lever last mentioned (**indirect motion**) starts the opposite piston on the reverse stroke, and this piston on reaching the end of the stroke opens the valve for the other piston to reverse.

Q.—Why is this arrangement made so?

A.—Because it is necessary to secure the reversal of the pumps, which could not be done in any other way as simple.

Q.—What gives the most trouble about a pump?

A.—Leaks, in one way or another.

Q.—How large a vacuum can be maintained conveniently in a suction pipe?

A.—About 28 inches by gauge.

Q.—How large a vacuum should there be?

A.—There should be 3 or 4 more inches of vacuum than feet of lift.

Q.—Suppose the suction-pipe was air-tight, and the vacuum was too low, what would generally be the cause?

A.—Leaky valves, or a leaky plunger, or a leaky stuffing-box, or a cracked cylinder.

FEED PUMPS

Q.—What is the proper size of the suction-pipe?

A.—The full size of the opening in the pump.

Q.—Would a **leaky water end** of a pump cause any unnecessary waste of steam?

A.—Yes; a pump would have to work at greater speed to keep the boiler supplied with water, than otherwise required, and this means a waste of steam.

Q.—Upon investigation what would you find to be the trouble?

A.—The water end improperly packed or no packing.

Q.—Suppose the pump used a good deal of steam and still went very slowly, causing great friction, to what would you lay the trouble?

A.—To the pump being packed too tightly.

Q.—How should the rings be fitted to prevent the friction?

A.—Cut the ring joints a little short to allow for expansion when wet or under pressure.

Q.—How would you try the pump to know that the packing is in good order?

A.—Close the delivery valve near the pump, let plunger make a stroke up and down. If then the pump stops of its own accord the plunger is well packed.

Q.—Is there any danger of bursting the pipe?

A.—No, for the pipe and valve should be able to withstand the pressure.

Q.—Is the pressure greater in the pipe than in the boiler; if so, what causes it?

A.—Yes, as there is more area in the steam cylinder.

Q.—How many valves has a **duplex pump**, and how many stories?

A.—The duplex pump has eight valves and two stories, four valves to each story, one story above the other. Larger pumps have more in proportion.

Q.—Where are the receiving valves located in a 3x4x6 inch duplex, also the discharge valves?

A.—The receiving or suction are the lower set and the discharge the upper set.

Q.—How would you partition off the valves in the water end of a double-acting pump?

A.—Place the partition between the suction valves only.

Q.—How many ports has each steam cylinder of a duplex pump? name them.

A.—There are five, namely, two steam, two exhaust and one outlet port to the atmosphere.

Q.—How wide are the steam, exhaust and outlet ports of a duplex pump 3x4x6 inches and how thick are the dividing ribs?

A.—The steam and exhaust ports are 7-16 of an inch, the outlet (in the center) is 9-16, and the ribs ½ inch each.

Q.—How long is the exhaust cavity in the middle of steam slide valve?

A.—It is 1 and 9-16 inches long.

Q.—How long are the two faces at each end of valve?

A.—They are each the same length as exhaust cavity.

Q.—How much lap has the valve at each end over the steam ports when in central position?

A.—Three-sixteenths of an inch.

Q.—How much lap has the exhaust?

A.—None; the valve just seals the two exhaust ports.

Q.—Why do large pumps have many small water valves?

A.—So the loss of water will not be so great in the rise and fall of the valves when the pump is working.

Q.—How is a vacuum created in a pump cylinder?

A.—There is not a real vacuum at any moment; the water follows the piston as fast as it moves, driven by the 14.7 lbs. per sq. in. pressure of the atmosphere.

Q.—To what height would the water follow the piston?

A.—To the height of 33 feet. At this point the column of water in the pipe would have the same weight as a column of the atmosphere with the same cross section area.

Q.—Why is a check valve placed near the boiler?

A.—To prevent the water in the boiler from

being forced back onto the pump at the end of each stroke.

Q.—What, if there were no check valve?

Ball Check Valve.

Horizontal Check Valve.

A.—The pump could be run, but the discharge valve would seat too hard and would wear out too soon.

Q.—What is done when the check valve needs repairing?

A.—The gate valve between the check valve and the boiler is closed.

Q.—Are angle check valves and vertical check valves used with boilers?

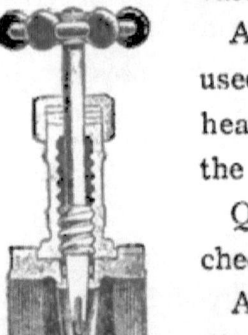
Sectional.
"Clip" Gate Valve

A.—Rarely; they are mostly used for connecting a steam heating system to the traps in the boiler room.

Q.—How much lift should a check valve have?

A.—Enough to give an area of opening, equal to the area of the feed pipe.

Q.—Is it well to give it a larger lift?

FEED PUMPS

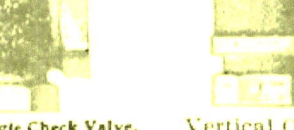

Angle Check Valve.   Vertical Check Valve.

A.—No; too much lift wears valve and seat.

Q.—What is the use of the pet cock?

A.—If the pump is in good order, the **pet cock** will show full stream at forcing and weak at suction. It shows tank or hydrant pressure both strokes, when receiving valve is held open by dirt, etc. It shows boiler pressure both strokes when check and discharge valves do not work properly.

Q.—What is the **air chamber's** duty?

A.—The elasticity of the air in it renders the pressure and flow practically uniform, notwithstanding the intermittent action of the force. It furnishes what in electricity would be called a constant potential service. It also renders the seating of the valves more even.

Q.—What are the features of a Fire Engine?

A.—A fire engine consists essentially of a pair of single-action suction and force pumps. The boilers are tubular, of sufficient capacity to work the pumps 500 strokes per minute. The working pressure of steam is usually 80 to 100 lbs. per square inch.

THE ARTESIAN PUMP

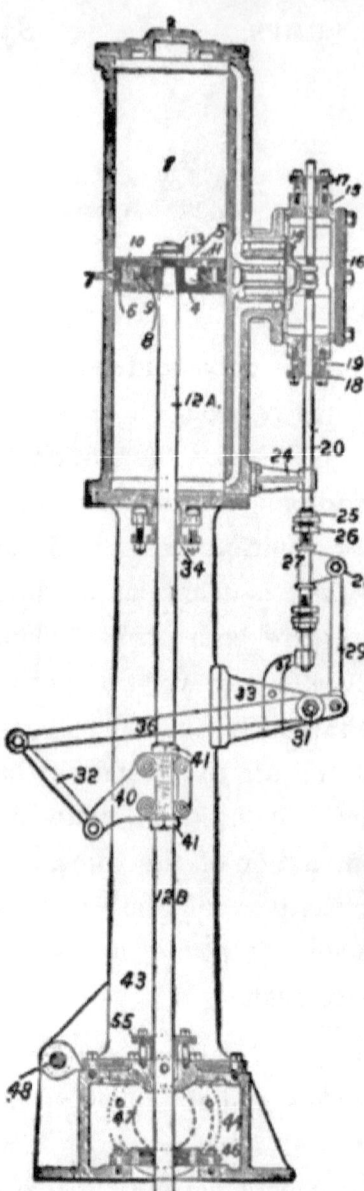

The engine part of the artesian pump shown in the cut, is a common vertical slide valve engine in its cylinder parts.

The figures indicate the parts as follows:
1. Steam Cylinder.
2. Steam Cylinder Head.
4. Steam Piston Head.
5. Follower Head.
6. Inside Piston Ring.
7. Outside Piston Rings
8. Adjusting Screw.
9. Jam Nut.
10. Adjusting Spring.
11. Cap Screws for Follower.
12 A, B. Piston Rod.
13. Brass Piston Rod Nuts.
14. Steam Slide Valve.
15. Steam Chest.
16. Steam Chest Cover.
17. Upper Stem Gland.
18. Lower Stem Gland.
19. Gland Studs.
20. Steam Valve Stem.
24. Stem Guide on Cylinder.
25. Brass Jam Nut.
26. Brass Split Nut.
27. Brass Tappet Head.
28. Tappet Head Bolt.
29. Stem Link.
31. Fulcrum Bolt.
32. Stem Guide.
33. Stand.
34. Piston Rod Gland.
36. Swinging Arm.
38. Crosshead Link.
40. Crosshead with Bolts.
41. Crosshead Jam Nuts.
46. Suction Flange in the Base.
47. Discharge Flange.
48. Hinge Bolt and Nut.
55. Base Stuffing Box Gland.

STEAM PUMP GOVERNOR

DESCRIPTION.—The upper wheel 1 in yoke is the lock nut. Turn it to the left; then turn lower wheel 2 to the right, which raises and opens the double steam valve 8; when partly open, open the throttle valve and start the steam pump. Now close angle valve 4 and open globe valve 5. This lets the main water pressure on the piston 6 and spring 3 in brass water cylinder 7. Now regulate by screwing up or down on wheel 2 until the water pressure gauge shows pressure desired to carry; then set in place by turning wheel 1 to the right until up against bottom end of the piston rod.

PUMP GOVERNOR

To OPERATE.—In starting or stopping the pump do it with the main steam throttle. Do not change the adjustment of the governor. In starting, close globe valve 5 and at the same time open angle valve 4. As soon as started, close angle valve 4, open globe valve 5 and pump will hold the pressure at which it is set.

PACKING GOVERNOR, ETC.—Pack the valve rod as light as you can and screw stuffing box nut down lightly with thumb and finger, just enough to stand the strain. Do not use wick packing, but some good sectional, square or round packing.

TO CLEAN AND OIL GOVERNOR.—Once a month run the pump by the throttle, shut off both valves 4 and 5, then open union 9, take off water cylinder cap 11, take out piston 6, also stem and steel spring 3, wipe out the cylinder 7, clean piston head 6, and oil them with some good oil that will not gum. If governor is kept clean and attention paid according to directions no trouble will arise.

TO CONNECT.—Place governor between the steam chest and throttle valve so it will stand plumb; connect bottom outlet flange or screw with steam pipe on steam chest, then connect the boiler pipe to inlet, placing throttle in most convenient place. Use short nipples so as to place governor as close to steam chest as possible.

TO CONNECT WATER PART.—Tap the discharge main or pipe, if horizontal, on the side for ¼ inch pipe, run pipe up about a foot higher than top of pipework of governor, then over to it and down and connect to quarter-inch valve on top of pipework over governor. If for two governors on pumps discharging into same main tap, the same as for single governor and run up and over between governors, then put on a "T," and run to right and left till over pipework above each governor and connect. If the pulsation of pump is noticed it can be avoided by partly closing globe valve 5. Never connect close to air chamber. Insert a short piece of pipe in drip "T" 12 at bottom of brass cylinder to reach the floor.

DOUBLE ACTION WATER PUMP WITH ROLLING VALVES
(See page 88)

A double action water pump, built in two pieces. A, the upper or main part, contains the delivery valves c,d, and the pump barrel c, which is made of a seamless drawn brass tube. (Fig. 2.)

The lower part contains the chamber B, to which the suction pipe is connected, and the suction valves a, b. It is bolted to A by bolts e, f.

The plunger, D, is provided with two reverse cup leathers. E, the plunger rod, passes through the stuffing box F.

The downward stroke of D opens the two valves a and c, while it closes b and d. The upward stroke acts in the opposite sense.

The deep well plunger, Fig. 1, consists of the brass pump cylinder A, the pump case B, the air barrel D, and the water pipe E connecting pump to stuffing box; c is the suction valve, b the delivery valve, f the suction strainer.

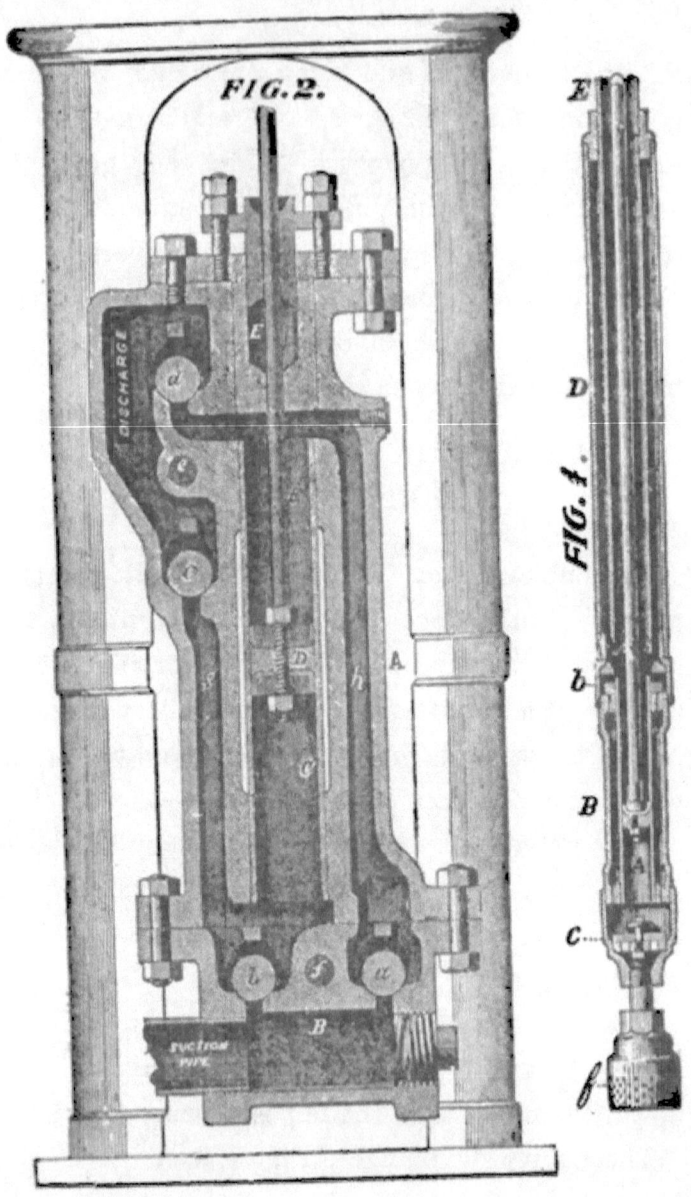

THE BOURDON STEAM SPRING GAUGE

VIEW OF INNER PARTS

DESCRIPTION.—A brazed, tempered-brass tube, bent in an almost complete circle, has the open end attached to one arm of a siphon pipe, while its closed end is fastened to a lever. The steam pressure on the water, in the pipe and tube, tends to straighten the tube or spring (by pressing more against the outer curve than the inner), moving the lever, the long arm of which turns, with its toothed arch, the hand of the dial, indicating the pressure per sq. inch of boiler.

Q.—Why is water kept in the spring and siphon?

A.—A direct contact with steam would take the temper out of the tube.

Q.—How is a vacuum gauge constructed?

A.—It has a lighter spring and it acts in the reversed sense as the atmospheric pressure tends to bend it more, the less pressure there is inside, or in other words, the greater the suction. The dial plate is graduated to register 30 inches of vacuum to equalize 15 lbs. of atmospheric pressure or a column of mercury of 30 inches.

Q.—Explain the compound ammonia gauge?

A.—It is the same style as a steam gauge, only the spring is of steel tubing and the graduation on the dial is to show both ways from zero mark. All figures above show pressure and all below show vacuum.

Q.—Why is the ammonia gauge spring made of steel instead of brass?

A.—Because the ammonia destroys brass, while steel is not affected by it.

Q.—Are there two springs in the compound ammonia gauge?

A.—No; it is so named because it shows either vacuum or pressure on the same dial with the same needle.

Q.—What is a duplex gauge?

A.—It has two springs and two needles, one showing the excess pressure, and the other train pipe pressure. They are used on locomotives only.

THE LUBRICATOR

Q.—Of what use is a lubricator?

A.—It supplies the valve, piston and cylinder with oil automatically after the drop feed is set.

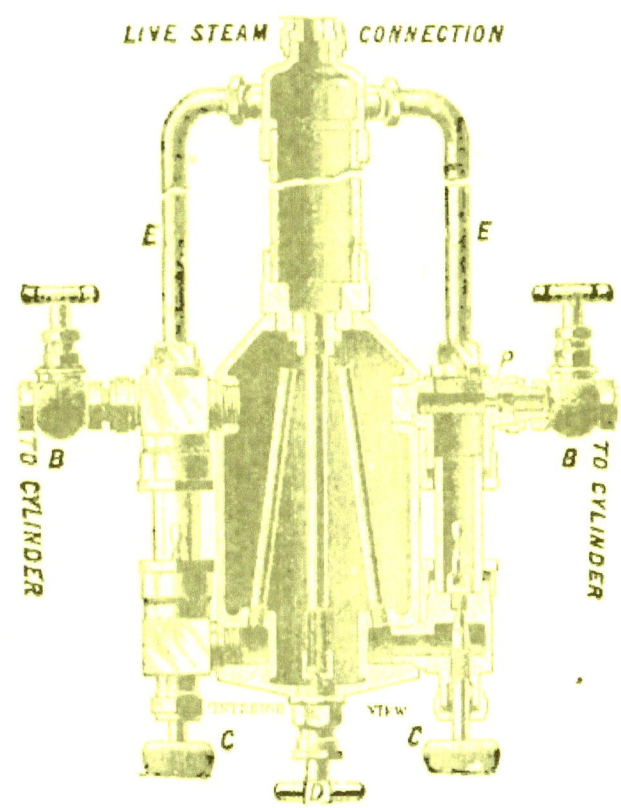

Q.—How many different styles of lubricators are there in use?

A.—There are three—single feed for stationary

engines, double feed for compound and locomotive engines, and triple feed for triple expansion, and locomotive engines and air pump.

Q.—How does a lubricator do its work?

A.—As seen in the cut, by the condensed water passing down the center water pipe from the condense chamber to the bottom of the oil reservoir, forcing the oil to the top and down the oil pipes to and through the feeder valves C, C. After passing the feeder valves, the oil floats up through the water in the sight feed glass and on reaching its surface is carried off horizontally through the choke plug (P) by steam from pipe E.

Q.—How is the lubricator **attached** to the system?

A.—Connect its top to the live steam pipe and the feed pipe further down.

Q.—What precaution should be had after attaching?

A.—The passages and connections of the lubricator should be blown out with steam.

Q.—How is a lubricator filled?

A.—Close all the feeder valves C, C, also the live steam connection, then open blow-out valve D, and fill through filler plug.

Q.—How is the feed stopped and started?

A.—By closing and opening the valves C, C.

Q.—Supposing the lubricator ran empty how would you refill it?

A.—Close feeder valve C and live steam connection; open blow-off D; open filler plug and as the water passes out, fill in with oil.

Q.—After filling what do you do?

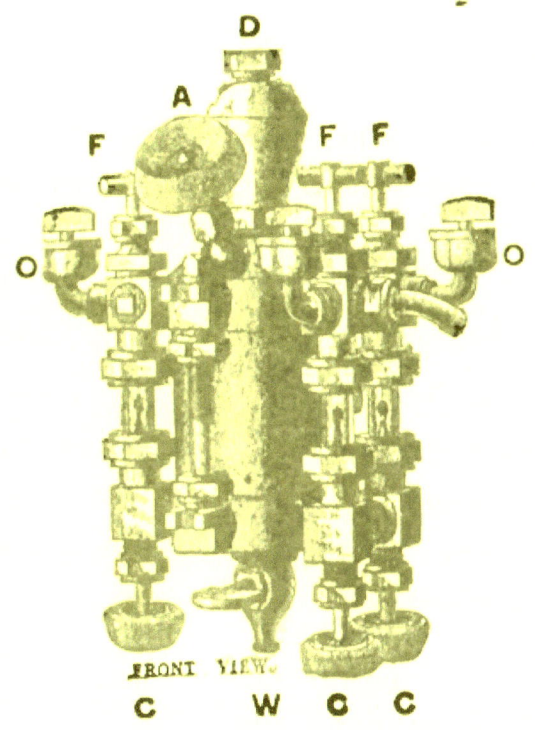

A.—First, close drip D, screw filler plug in tight, open live steam connection fully, then regulate oil flow with valves C, C.

Q.—Are the valves B, B ever closed?

A.—No, except when a feed glass is broken.

Q.—What is to be done, when a glass breaks?

A.—Close valves C, C and B, B; unscrew plug of **top bracket, loosen packing nuts and remove old** glass. Insert the new glass, and fasten nuts and valves, etc.

Q.—Is it well to reuse old gaskets?

A.—It is not.

Q.—Is there any difference between the single, double and triple lubricators?

A.—Not in principle or operation.

Q.—Do all lubricators work alike?

A.—No, the down feed lubricator dispenses with the water in the feed glass.

Q.—How much oil should be fed through a lubricator for an engine working heavily?

A.—That depends, of course, on the quality of the oil, and also, of course, on the condition of the engine. For heavy work 2-5 drops a minute, for light work 1-4 drops.

Q.—What care should be taken in filling a lubricator?

A.—No foreign matter must be allowed to get in. The opening in the feed nozzle is so small that almost anything would clog it.

Q.—How large is the opening?

A.—It is 3-32 of an inch.

Q.—How is the feed nozzle cleared of clogging matter?

A.—By shutting off the live steam connection, opening the blow-off and then opening the feed valve to allow the back pressure to pass through the opening.

Q.—How is the choke plug cleared of clogging matter?

A.—In the same way. The back pressure will force the matter into the feed glass.

Q.—How large is the opening in the choke plug?

A.—It is 3-64 of an inch.

Q.—How can it be decided whether this opening is of the right size?

A.—Start the engine. Then regulate the oil feed in the glass, counting the number of drops in one minute. Then shut the throttle of the engine and notice quickly whether the number of drops changes. The number will not change if the opening is of the proper size.

Q.—Is it harmful to use more oil than needed?

A.—Yes. It clogs up the exhaust pipe of the engine, decreasing its opening. The increase in pressure necessary for exhausting through a clogged exhaust pipe means larger coal consumption. Besides, it is a waste of oil.

Q.—Is it well to feed both valve-oil and engine-oil through a lubricator?

A.—No. The mixing impairs the lubricating properties of the oils. Only valve-oil should be used for engine cylinders.

THE ENGINE

THE COMMON SLIDE VALVE

DESCRIPTION.—When the piston is at either end of the cylinder, the steam port at that end is open a fraction of an inch (lead); the steam enters

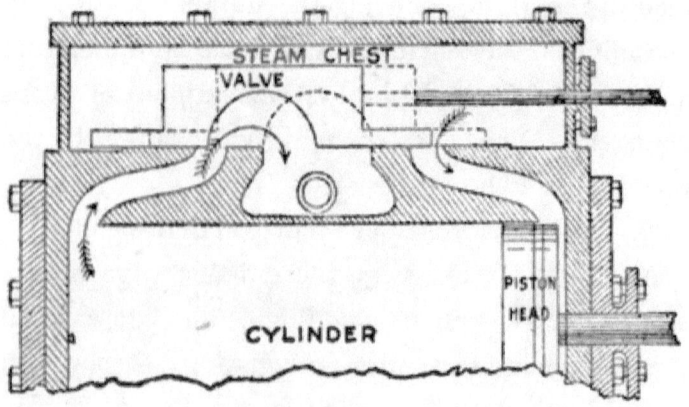

and starts the piston on its travel, the port opening wide and admitting the steam freely. The valve travels in the opposite direction. When the piston has traveled ¼ or so of its stroke (according to the lap on the valve) the slide closes the steam port, so that during the remainder of the stroke no more steam enters on that end of the cylinder. The steam present expands, therefore, as long as the piston keeps moving in the same direction. At the moment when the piston reaches the other end of the cylinder, the steam port there opens

THE ENGINE

slightly (lead), the entering steam pushes the piston back and the expanded steam on the other end of the piston escapes through the exhaust cavity of the valve, which at that moment connects the steam port with the exhaust port, and disconnects them again when only enough steam is left to serve as a compression at the point from which we started. The operation is the same at both ends of the cylinder.

Q.—How would you proceed to set a common slide valve?

A.—See that valve covers both steam ports equally, the crank pin at dead center, heavy side of eccentric up or at right angle with the crank pin, rocker arm plumb, at center of travel, and all connections close fit; move the eccentric around on the shaft in direction engine is to run, until the valve has proper lead, say 1-16 of an inch, then tighten eccentric with set screws, turn crank pin to opposite dead center and see how the lead is at the opposite port. If equal the valve is set; if not, divide the difference by adjusting the length of the valve rod and readjusting the eccentric.

Q.—Is the common slide much used?

A.—Yes, in the cheaper forms of steam-engines, compressed-air engines, in some air-compressors, and in some compressed-air ice machines.

Q.—How is the length of the valve stem and of the eccentric rod found?

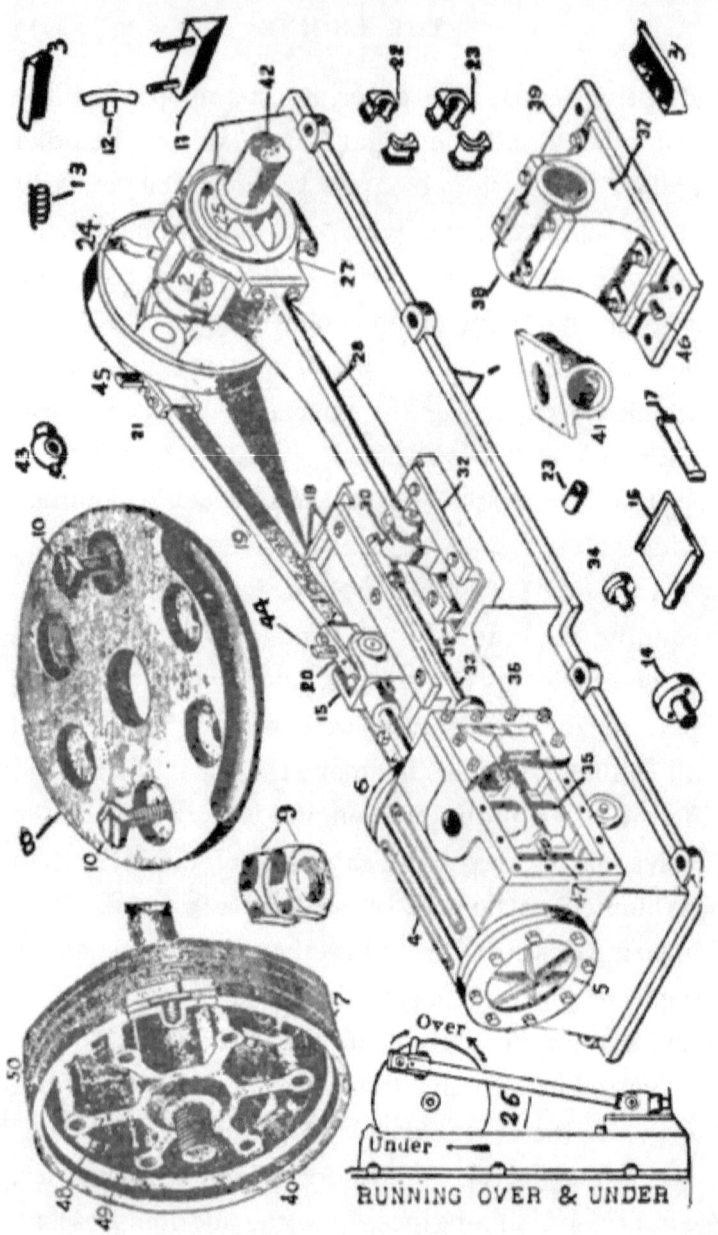

Explanations of Sectional Cut of Common Slide Valve
AS SHOWN ON OPPOSITE PAGE

LIST OF NUMBERS AND CORRESPONDING PARTS

1. Bed Plate (frame).
2. Main Bearing Cap.
3. Main Bearing Quarter Box-Brass.
4. Cylinder.
5. Cylinder Head. Back.
6. "   "   Front.
7. Piston, Spider and Rings.
8. Follower Head Plate.
9. Follower Nut and Jamb Nut
10. Follower Bolts.
11. Main Bearing Quarter Box Wedge.
12. Piston Packing Shoe.
13. Piston Packing Spring.
14. Piston Rod Gland.
15. Piston Cross Head.
16. Piston Cross Head Shoe.
17. "   "   "   Gib
18. "   "   "   Guide Cap Plates.
19. Connecting Rod.
20. "   Strap, Cross Head End.
21. "   "   Crank End.
22. "   "   Cross Head Boxes.
23. "   "   Crank Pin Boxes.
24. Crank (disc).
25. Eccentric.
26. Engine Showing Over or Under Running.
27. Eccentric Straps.
28. "   Rod.
29. "   "   Bushing.
30. Valve Cross Head.
31. "   "   "   . Guide Caps
32. Valve Stem.
33. "   "   Gland.
34.
35. Main Steam Side Valve.
36. Steam Chest Cover.
37. Outboard Bearing or Tail Block.
38. "   "   "   Cap.
39. "   "   "   Sole Plate.
40. Head of Piston.
41. Exhaust Elbow.
42. Main Shaft.
43. Eccentric Rod Head.
44. Rod Key Cross Head End.
45. "   "   Crank End.
46. Adjusting Stud of Tail Block.
47. Adjusting Nuts of Valve.
48. Spider.
49. Bull Ring.
50. Piston Rings.

A.—Place the valve in the middle of its seat. In this position the rock arm is vertical. The distance A between a vertical line drawn through the center of the rock shaft and a vertical line through the center of the valve is the length of the valve stem. The distance B between the central

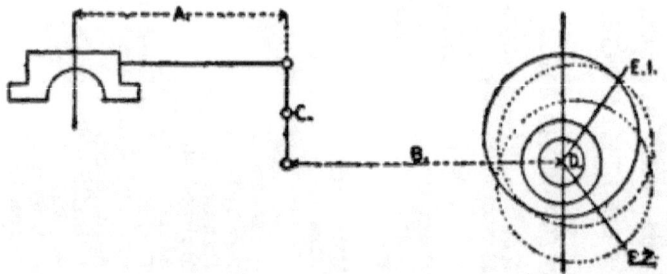

vertical of the rock shaft and a vertical drawn through the center of the main shaft is the length of the eccentric rod.

Q.—How is a single eccentric engine **reversed**?

A.—Remove the valve chest cover, measure lap and lead, loosen the eccentric and move it around the shaft to a position where the valve will show exactly the same as before. Then fasten the eccentric with the set screws in position and cover the valve.

In the cut above position E1 shows the engine running over forward. By bringing the eccentric into position E2, the engine is reversed, running under backward. Further adjustments can then be made by repeatedly trying the engine from one dead point to the other.

Q.—How would you reverse the motion of a common slide valve engine?

A.—Set the crank pin on dead center, remove the valve chest cover and notice the amount of lead at steam edge of valve. Then loosen the set screw or key of the eccentric and turn it around on the shaft in the same way it has been running until the valve has reached the end of its travel; keep moving the eccentric until it has the same lead as before, then tighten the set-screw.

The dotted circles in the cut on page 100 indicate the positions of eccentric other than plumb.

Q.—How is a single eccentric slide valve rocker arm engine converted into a reversible engine?

A.—If the valve has neither lap nor lead, it can be done by putting another wrist pin in the rocker arm, above the rock shaft center. But if the valve has outside lap, then another eccentric must be put on, and both eccentric rods hooked on—in their turn, according as the engine is to run—to the same wrist pin placed below the center of the rock shaft.

Another way is to use the link motion, as in a locomotive engine.

Q.—Which leads, the crank pin or the eccentric?

A.—The eccentric. (See page 116.)

Q.—At what angle does the highest-pitch line of an eccentric lie to the level of the crank pin at dead center?

A.—About 120 degrees.

Q.—What is initial pressure?

A.—Initial pressure signifies the pressure present in the cylinder of an engine at the beginning of its stroke.

Q.—Explain terminal pressure?

A.—Terminal pressure is in the cylinder of an engine at the end of the stroke of piston if the exhaust valve does not open until the stroke is finished.

Q.—What is meant by wire-drawn steam?

A.—The operation of reducing the steam pressure between the boiler and the cylinder.

LEAD AND LAP

Q.—What is pre-admission or lead of an engine?

A.—The amount of steam port opening just before the end (or: at the very beginning) of either stroke of the piston.

Q.—How would you give lead to a valve?

A.—By moving the eccentric in the direction that the engine is to run. (See page 100.)

Q.—How would you alter a valve to cut the steam off at a given part of the stroke?

A.—As the case demands add lap to or take off from the steam edge of valve.

Q.—What is lap?

A.—The distance that the valve laps over the steam-points, when in mid-position.

THE ENGINE

Q.—Is there any lap on the exhaust edges of valve?

A.—Yes. They serve to delay and shorten the exhaust and thus to increase compression.

Q.—Have all engines exhaust lap?

A.—Practically yes, and they have the more lap, the shorter and quicker their travel.

Q.—Why is lap given to a steam valve?

A.—So the port will close before the piston reaches the end of the stroke and make the steam to work by its own expansion.

Q.—What is meant by "¼ cut-off"?

A.—In an engine with a 24″ piston stroke it would mean that the steam port is closed when the piston has traveled 6 inches, or the *first* quarter of its stroke.

Q.—How many expansions are there in our case?

A.—At the quarter stroke, one expansion; at the half stroke, two expansions; at the three quarter stroke, three; and at the full stroke, four.

Q.—How do you decide the proper amount of steam lap on a slide valve?

A.—The length of the piston stroke, minus the part of stroke before the cut-off, is divided by the whole length; extract square root of the quotient. Multiply this square root by one-half the length of the stroke of the valve, and from the product take one-half the lead (if any), and the remainder will be the amount of lap required.

Example: Given a 48″ stroke, travel of valve 6″, to cut off at half stroke, no lead. Half stroke equals 24″. $24 \div 48 = .50$. $\sqrt[2]{.50} = .707$. $.707 \times 3 = 2.121″$. Ans. (Sq. root, see page 240.)

Q.—Should the lead and compression in the valve of a vertical engine be the same on both ends of the cylinder?

A.—No; the lead and compression at the lower end of the cylinder should be greater, to make up (compensate) for the weight and momentum of the piston, crosshead and connecting rod.

Q.—What is the difference between movable and fixed expansion?

A.—Movable expansion is by separate gearing or valves, and fixed expansion is by lap of slide valve.

Q.—Give another name for the expansion valve for cut-off?

A.—The link motion. (See page 144.)

COMPOUND ENGINES

Q.—Explain the compound engine?

A.—A compound engine is one that has two or more cylinders following in regular order having increasing diameters so arranged that the exhaust steam from the first or high pressure cylinder is exhausted into the second or larger cylinder to do work before escaping to the condenser.

Q.—Explain the special advantages by compounding?

A.—First—Compounding enables the fullest advantage to be taken of the expansive power of very high pressure of steam. Second—The ease with which it may be adapted to work on one or more cranks, thereby reducing the excessive variation of strain which occurs in a single high pressure engine.

Q.—Classify compound engines?

A.—First—Where the pistons of both cylinders commence the stroke at the same time. Second—Those which exhaust from one cylinder before the next cylinder is ready to receive it, in which case the steam is retained for a portion of the stroke in a chamber or receiver between the two cylinders.

Q.—What kind of engines would you call them?

A.—**Receiver engines**.

Q.—What can you say about triple and quadruple expansion engines?

A.—The principles which govern the compound are the same in the **triple and quadruple expansion engines**.

Q.—What is **a cross compound**?

A.—It is two separate engines side by side connected to one shaft, one being a high pressure and the other a low pressure. (See cut page 10.)

Q.—What is meant by the **tandem compound engine**?

A.—It is where the high and low pressure cylinders and engine frame are one behind the other (one following the other).

Q.—About how high should the steam pressure be to run a tandem or cross compound engine with economy?

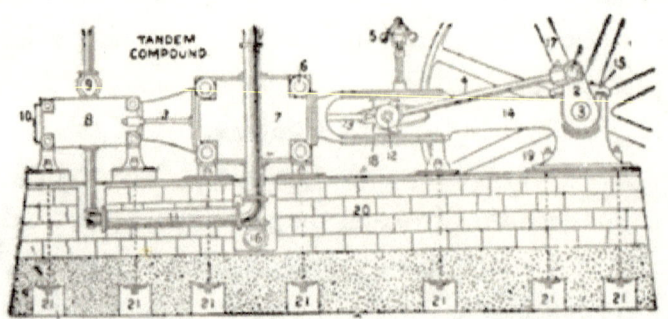

NAMES OF PARTS OF TANDEM COMPOUND ENGINE

1. Crank Pin.
2. Crank.
3. Crank Shaft.
4. Main Rod.
5. Governor (fly ball).
6. Valve Chest Covers
7. Low Pressure Cylinder.
8. High Pressure Cylinder.
9. Globe Valve.
10. Cylinder Head.
11. High Pressure Exhaust Pipe.
12. Cross Head.
13. Piston Rod.
14. Engine Frame.
15. Main Cap.
16. Low Pressure Exhaust.
17. Fly Wheel.
18. Guides.
19. Foundation Bolts.
20. Engine Foundation.
21. Bottom Foundation Bolts.

A.—The boiler pressure should be from 110 to 125 lbs. pressure to the square inch to work well. For triple expansion about 180 lbs., and so on in proportion.

Q.—Why is such high pressure carried for compound or triple expansion engines?

A.—So the last low pressure cylinder will be

able to do work from the expansion of steam that first entered the high pressure cylinder.

For calculating the H. P. of a compound engine, see page 150.

THE CORLISS ELECTRIC ENGINE STOP

DESCRIPTION. The steam valve K is held closed by the electro-magnet O and armature G, as long as the lever H is in upright position. When the circuit P P is closed, the upper end of lever H is released and steam can open the valve, forcing H into the position indicated by dotted lines. The steam is thus admitted into the cylinder A, closing the vertical check valve J and forcing the piston C upward. The end of the piston rod C engages the clamp D, attached to the side rod M, raising the governor balls N N to their highest position. Then check valve L closes, holding the governor balls in their highest position. Thus the grab hooks are prevented from opening the main valve.

The electric circuit P P is closed at will by pressing a button conveniently located, or it may be closed automatically at any desired limit of speed by means of the SPEED LIMIT STOP ATTACHMENT.

This device consists of two adjustable points F F, electrically connected and so placed that the

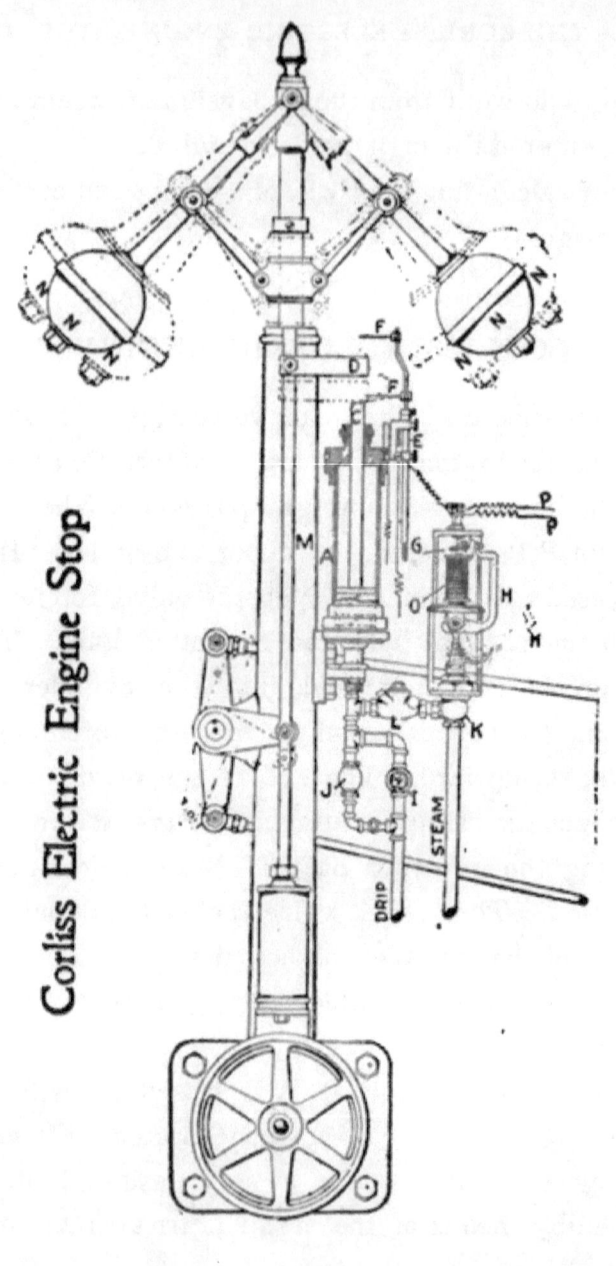

Corliss Electric Engine Stop

clamp D is brought in contact with one of them at the highest, and with the other at the lowest, desired limit of speed. This contact closes the circuit, the lever H is released, etc., as above described.

To put the engine stop in working order again, the lever H is replaced in the upright position, the drip valve I is opened to allow the governor balls to resume their normal position, and the valves are set in the proper position by rocking the wrist plate backward and forward.

THE HOT AIR PUMPING ENGINE
(See Sectional View on next page.)

The air heated in F expands, driving the power piston D upward. The compression piston C moves downward at the same time, driving the cold air through H into F, thus increasing the pressure until its stroke is completed. Then the increased pressure forces the compression piston C up, passing back from F to C through the regenerator H, where the heat is absorbed by the regenerator plates. By this cooling the pressure is lowered to its minimum, the power piston descends and compression begins again. The air passing from C to F through H, reabsorbs heat from the regenerator plates, which helps to augment the pressure in F. (Pump, see page 88.)

VIEW OF COMPRESSION ENGINE

DESCRIPTION

A.—Compression Cylinder.
B.—Power Cylinder.
C.—Compression Piston.
D.—Power Piston.
E.—Cooler.
F.—Heater.
G.—Telescope.
H.—Regenerator.
II—Cranks.
JJ.—Connecting Rods.
KK.—Piston Packings. (Leather.)
L.—Check Valve, placed at back of compression cylinder, but shown at side on cut.
M.—Pump Primer.
N.—Blow-off Cock.
OO.—Knuckles.
PP.—Heater Bolts.
R.—R'g'n'r't'r Bon'et.
SS.—P'mp V've B'net.
T.—Water-Jacket, to protect packing from heat.
UU.—Pump Buckets.
V.—Pump Gland.

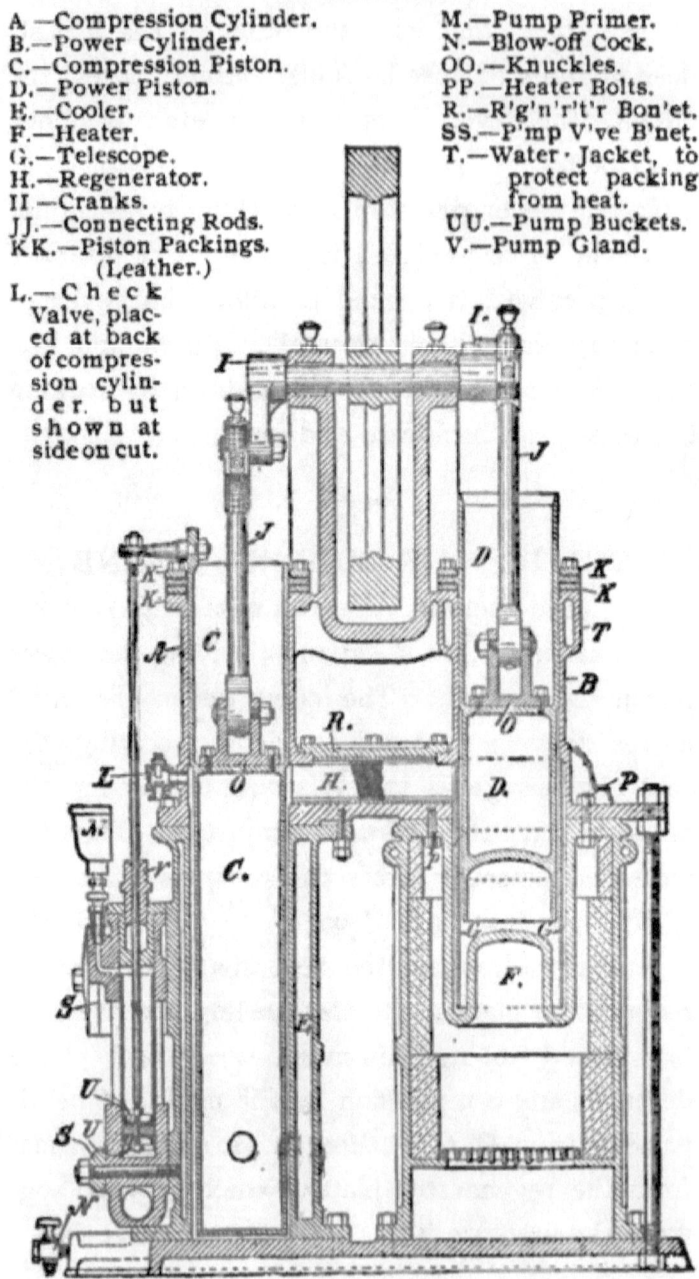

OILING DEVICES 111

The two ends of each **connecting rod** are connected by a tube A, in which is fitted a rod B, extending from the upper to the lower brass, and so arranged that one key, E, at once takes up the lost motion *on both brasses* C, D. The key is nicely adjusted by the nuts F, G.

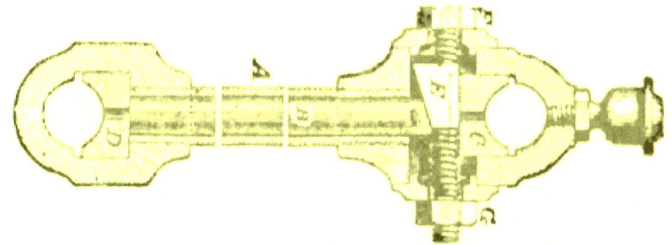

The pump plunger U U moves up and down with the compression piston C, to the top of which the plunger rod is connected.

SIGHT FEED AND OILING DEVICES FOR ENGINE AND MACHINERY BEARINGS

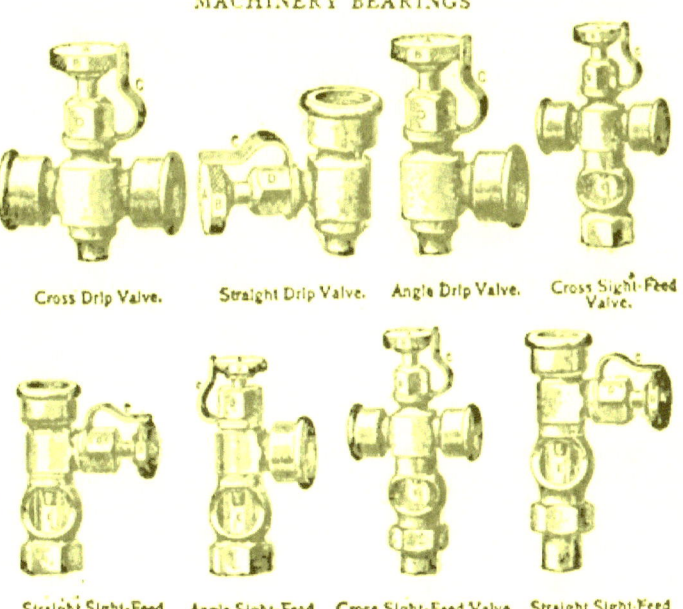

QUESTIONS AND ANSWERS

A.—Regulating valve.
B —One of the flat places to hold spring C.
D.—Packing nut.
E.—Light glass.

Angle Sight-Feed Valve with Union.

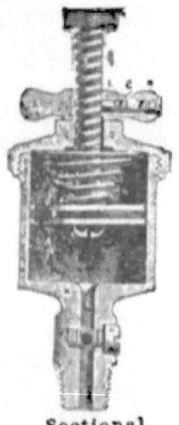

Sectional Grease Cup.

WIPER CUPS

Adjustable Wiper Cup for Wick.   Adjustable Wiper Cup with Elbow Shank.   Adjustable Wiper Cup for Crank Pin.   Adjustable Plain Wiper Cup with Elbow Shank.

Plain Wiper Cup.   Horizontal Wick Wiper Cup.   Oil Cup Wiper Tip.   Drip Trough.

CONDENSERS

Q.—What is a **condenser** as applied to an engine?

A.—It is a part of the low pressure engine and is a receptacle into which the exhaust enters and is there condensed.

Q.—What are the principles which distinguish a **high** from a **low** pressure engine?

A.—The high pressure is over 40 lbs. and exhausts into the atmosphere, while the low pressure is below 40 lbs. and exhausts into the condenser.

Q.—What is the object of a condenser as applied to an engine?

A.—It saves a large mass of pure hot water for the boiler, and it maintains a constant vacuum in front of the piston.

Q.—Does this vacuum aid the steam in moving the piston?

A.—It does, to the amount of half the vacuum gauge pressure.

Q.—How is a vacuum maintained in a condenser for a compound condenser engine?

A.—By the steam used being constantly condensed by the cold water or cold tubes and the air pump continually clearing out the condenser.

Q.—How does the condensed or used steam form a vacuum?

A.—Because a cubic foot of steam at atmospheric pressure shrinks into about 1 cubic inch of water.

Q.—What is a **surface condenser?**

A.—It is a chamber or receiver for the exhaust steam, through which pass brass tubes, carrying the cold water which is supplied usually by a circulating pump.

Q.—Why is it called a surface condenser?

A.—Because the exhaust steam is condensed by the surfaces of cold water tubes and then removed, together with the air and vapor, by means of an air pump.

Q.—Where is a surface condenser most desirable?

A.—Where condensed steam is used for feeding boilers and distilled water is used for making pure ice.

Q.—Explain the **jet condenser**?

A.—It consists of a chamber in which the exhausted steam passes through a spray or jet of cold water. The steam, being condensed, falls with the injection water into a hot well, and from there it is pumped out.

Q.—Is there a valve between the pump and a jet condenser? If so, of what use is it, and what trouble is it likely to give?

A.—A foot valve is sometimes placed between the pump and condenser. Its object is to close the condensing chamber on the down stroke of the bucket plunger of a single acting pump, allowing a partial vacuum to be maintained in case of the failure of a valve on the bucket. The trouble likely is the same that affects other water valves under the same conditions.

Q.—Is it well to start the jet condenser together with the engine?

A.—It should be started before the engine begins to revolve, and it must be started very gradually.

Q.—What is meant by a **vacuum**?

A.—A space void of matter.

Q.—Can a perfect vacuum be obtained?

A.—No, but a compound condensing engine exhausts about 14¼ out of 15 lbs. of atm. pressure (indicated at 28½ inches on vacuum gauge, 2 inches representing one pound).

Q.—Suppose a vacuum gauge indicates 26 inches, how many pounds would it represent?

A.—Thirteen pounds.

Q.—What does 13 lbs. or 26 inches vacuum signify to an engineer of steam?

A.—That he may work his steam down to 4 lbs. before it exhausts, as the condenser utilizes 13 of the 15 lbs. atmospheric pressure. Without the condenser the engine would work high pressure.

Q.—What is meant by the term "**back pressure**"?

A.—It is the pressure that hinders the piston, equal to the difference between a perfect vacuum and what the gauge reads.

Q.—Is it possible to run both cylinders of a compound engine with high pressure?

A.—It is a very rare thing to feed a low-pressure cylinder with live steam. Few of them would stand a high pressure.

THE ECCENTRIC

Q.—Explain the length, throw or half the stroke of engine (crank)?

A.—It is the distance between the center of crank pin and center of shaft.

Q.—Explain the eccentric, also its throw?

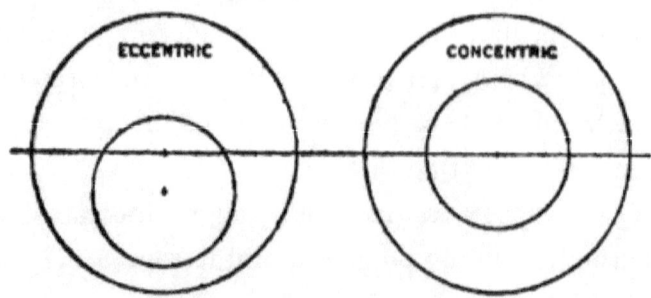

A.—It is anything out of center, or not concentric. It serves as a substitute for a crank. The throw or stroke is the distance between the centers.

Q.—Explain what is understood by the travel of a slide valve?

A.—It is twice the throw of the eccentric.

Q.—If the eccentric was made a half inch larger or smaller and the throw left the same, would it affect the travel of the valve?

A.—No; it only affects the straps.

Q.—Why not?

A.—Because the throw or stroke is the only point that regulates the travel of valve.

Q.—Explain the meaning of direct and indirect valve motion?

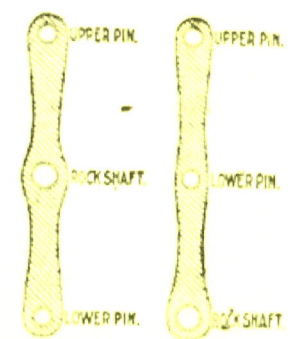

A.—When the two connections are on the same side of the rock shaft, they move in the same sense (direct). When they are on opposite sides of the rock shaft, they move in opposite senses (compound or indirect).

Q.—Can you set a slipped eccentric without moving steam chest cover, and how?

A.—Yes. Open cylinder cocks, roll crank pin over in the direction engine runs till the pin is on "dead center," then open throttle slightly. Roll the eccentric forward in the direction the engine runs until steam escapes from cylinder cock at the end where the valve should begin to open, screw the eccentric fast to the shaft, roll the crank over to the other "dead center" and see if steam escapes at the opposite end of cylinder; if so the engine is ready to run until an opportunity occurs to open the valve chest and examine the valve and set properly.

DEAD CENTER

A simple way of finding the exact "dead center" of an engine is as follows:

Place a stationary rest close to the rim's edge of the fly wheel or crank disc, at the side furthest away from the crosshead; roll the crank pin over, until the crosshead is within one-half inch of the end of its travel in one direction; make a well-defined mark on the guide at the end of the crosshead; then lay a square across the stationary rest and with chalk and a scriber mark the rim's edge of the fly wheel or disc.

Then roll the crank pin in the same direction past the center until the same end of the crosshead comes to the same mark on the guide, then lay a square on the stationary rest and scribe another line on the rim's edge of fly wheel or disc.

Then take a pair of dividers and find the middle between the two points just marked This middle point will correspond exactly with the "dead center" of the engine. Mark the point with a center punch.

Then roll the engine over, until the crosshead comes to the opposite end of the slides, and proceed as before to find the "dead center" at the other end of the stroke.

Another way is by the use of an adjustable spirit level. The level is adjusted to the guides and then the crank pin strap is adjusted to that level.

ENGINE POUNDING

Q.—Can you describe the common causes of an engine pounding?

A.—Yes.

Q.—Name them?

A.—Crank pin not being square (at right angles, L) with the crank, caused by faulty workmanship, an engine being out of square, lost motion in crank, in crosshead pin or journal boxes, leaky piston rings, unbalanced valve, crank disc and pulley wheels, valve not being properly set, poor lubrication, loose piston-head, water in cylinder, and many others.

Q.—Can you tell whether a crank pin is out of square?

A.—Yes.

Q.—In what way?

A.—The slightest variation can be found by a good spirit level, which is used as follows: First, disconnect the main rod from the crosshead, then key the rod to the crank pin so the rod can turn without moving sidewise; then place the rod in a position to move easily as the crank is turned.

Fasten a spirit level to the rod with a clamp in line with the shaft (right angles with rod). When the crank is turned the bulb in level should not change. If it does the pin is not square with disc nor parallel with shaft.

Q.—Does it make any difference if the shaft is not level when testing crank pin?

A.—No; the point that the bulb in the spirit level moves to, when clamped to the rod, is the point where it should stay the full revolution.

Q.—If the crank pin is not properly set, how will the bulb show?

A.—It would move, and the more the pin is out of square, the more the bulb will move.

LINING, LEVELING AND SQUARING AN ENGINE SHAFT

Q.—Name handy tools for lining engines and taking measurements?

A. -Spirit level, light chalk line, calipers, rule, square, slotted piece of wood for end of cylinder to hold line, and a tram.

Q.—How would you line an engine, also square and level a shaft?

A.—First, disconnect and remove all parts from crank pin to back cylinder head, then bolt a slotted stick to farthest end of cylinder from crank pin, to which attach a string and pass it over the point, thence through the cylinder to end of bed plate and fasten so it can be adjusted to the center.

Q.—How is the adjusting done?

A.—With inside calipers.

Q.—From where do you line?

A.—From the two counter-bores.

Q.—Why center from counter-bores and not from the regular bore of the cylinder?

A.—Because they are not worn, while the regular bore is.

Q.—How would you square the shaft of an engine?

A.—Move the crank pin both ways, forward and back, above the center line of cylinder. If the center line intersects the crank pin both times at the same point, then the latter is square with the cylinder. If not, shift the outbearing of shaft.

Q.—How would you level the shaft?

A.—Drop a plumb line below and near center line and in front of the center of shaft, then try the pin at top and bottom—half strokes—same as in squaring.

Q.—How would you know if shaft was in line with center of cylinder or proper height?

A.—By placing square against disc (or crank) face and bringing up to line.

Q.—How is the proper length of main rod found,

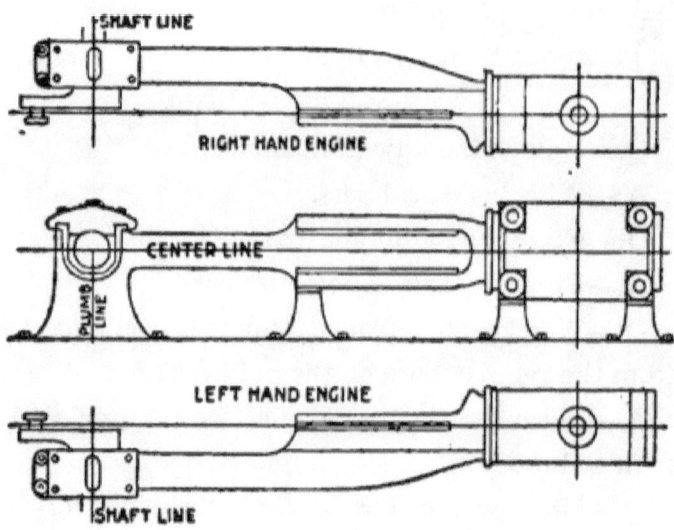

also the clearance between the piston and cylinder head when engine is at either dead center?

A.—By first finding the striking points, pushing the piston to one end of cylinder, then to the other, marking the crosshead and guide. After this is done find the full stroke of engine by measuring from center of shaft to center of crank pin. The distance found is one-half of stroke. The difference between full stroke and the two

LINING, LEVELING AND SQUARING 123

striking points is full clearance for both ends. After this is known move the crosshead (with piston attached) back from striking point one-half of full clearance, which will give an equal clearance for both ends, viz.: Distance between striking points is $17\frac{1}{2}$ inches, stroke of engine 16 inches, full clearance $1\frac{1}{2}$ inches; the half will be $\frac{3}{4}$ inch clearance at each end of cylinder. Then place the crank pin on same dead center at which the crosshead is placed and measure the length of rod with a tram from center of crank pin to center of crosshead (wrist) pin.

Q.—How are the **guides** put in line?

A.—For level, lay a straight edge across the two guides and caliper **between** it and the center line the whole length of the guides. For proper alignment across, caliper between the guide edges and the center line the whole length.

Q.—What do you do, after the guides are properly lined?

A.—Remove the line. Place the piston in the cylinder and place the crosshead on the piston, keying it on, or screwing it on as the case may be. Then line the piston rod by the guides for level and sideways at both ends of guides.

Q.—What do you do when the cylinder is worn so that the piston center is out of line.

A.—Put shims of tin under the spider between the lugs and bull ring, until the piston center is central, then adjust the packing rings by the setting out or tension springs.

Q.—How do you line the **crank pin** with cylinder?

A.—Place the connecting rod on the crank pin, and key up the brasses until they hug the crank pin snugly. Then move the crank pin to one of the dead points, and measure with inside calipers how far the side of the brasses on the opposite end of the rod is from the guide. Then move the crank to the other dead center and measure the distance on the other end of the guide. If these two distances are the same, the crank pin is perfectly in line.

AUTOMATIC SHAFT GOVERNOR FOR SIDE CRANK ENGINE

DESCRIPTION: In following cut, A indicates the hole in the tripod (a seat or instrument with three feet or arms) B, through which the engine shaft passes. The eccentric C is hung to the long arm of the tripod B by a stud and secured by a screw with washer. The hole through the eccentric is much larger than the shaft, which permits the center of the eccentric to shift across the shaft, thus varying its throw. The eccentric is supported and guided by a gibbed rebate fitting over a projecting lip on the tripod. The dead wheel D is fitted loosely on the hub of the tripod so that it may remain stationary while the tripod turns within it. The weights E, E are pivoted to the dead wheel by pins, fastened with set screws, and are connected with studs to the short arms of the tripod

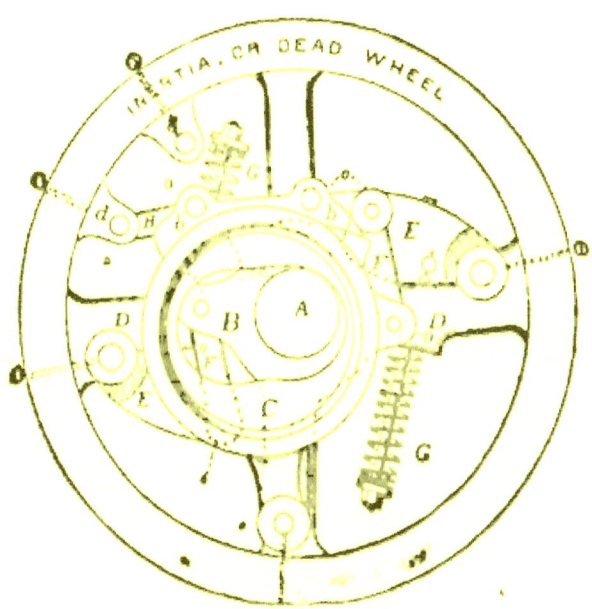

by the weight links F, F. The springs G, G are pivoted to the weights by their rods, and rest upon lugs on the arms of the dead wheel. They are precisely alike, acting together as one. The weights and weight links, as well as the springs, are duplicated only to secure more perfect balance of the governor. The eccentric link H connects the eccentric with the rim of the dead wheel by pins, which are made tapering to provide means for taking up the wear. O, O, O, O, O represent oil holes in dead wheel.

To Set the Valve: The governor should set on the shaft so that the center of the long arm of the tripod, to which the eccentric arm is attached, is on the other side of the main shaft, directly opposite the crank pin, and keyed in that position.

The steam valve that the governor controls should be adjusted to uncover the ports an equal distance on each end and should be set (while the weights are blocked out) as close to the rim of the dead wheel as the set-screw in the outer side of one of the weights will allow them to go.

Place the engine crank pin on the inner dead center and allow the valve to just cover the steam port at the outer end of the cylinder; that is, the outer end (edge) of the valve being line and line with the outside edge of the outer port. Turn the crank pin (or engine) in the direction it is to run (over or under) to the outer center, and the inner end of the valve should correspond in like manner with the outside edge of the inner port. If it does not, equalize the difference by the nuts on the valve rod.

Roll the engine to the inner center again to be certain that the adjustment is right, and when this is accomplished the valve is correctly set.

The governor should be so placed on the shaft that the eccentric is exactly in line with the valve crosshead pin and does not touch the side of the main shaft bearing.

MANAGEMENT AND CARE: All movements from the steam valve to the governor parts should be free, smooth, and without lost motion. To have the governor in order keep it clean and all pins and bearings well oiled.

The cut shows five oil holes through the rim of the dead wheel, marked O. There are also two in the hub. These are closed with plugs which are removed when oiling. Use good oil and oil frequently. Should the governor work irregularly or fail to control the engine the cause will usually be found in some dry joint or place that binds.

When there is a set screw on the inner side of the weight it is to limit the travel toward the hub, and should never be removed or disturbed. The governor key should fit closely on the sides of the keyway, but never on the top and bottom, to avoid springing the tripod hub and causing the dead wheel to bind.

If governor should become gummed from bad oil, take out springs, first carefully measuring their length in position. This precaution is necessary so they may be replaced exactly in the same position. Take out one at a time to avoid disarrangement. Clean with kerosene, etc. Before tension is again put on the springs, move the dead wheel back and forth to see that there is no binding in any of the working parts.

To CHANGE SPEED by changing the tension of

the springs: To run faster, tighten; to run slower, loosen. Or move the weights. Never tighten the springs down so that their spirals touch each other. Never tighten or loosen the spring more than one inch beyond its set tension. If greater speed be desired than can be obtained from the springs and weights furnished with the governor, others should be ordered from engine builder. Be sure to mention speed desired.

REVERSING governor and engine to run over or under: Remove the set-screw from the outer side of the weight. Disconnect the eccentric link H from the eccentric and from the rim of the dead wheel by removing the pins from the lugs b and d. Take hold of the rim of the dead wheel with a monkey-wrench and pull it around on the shaft as far as it will go, in the direction the engine is to run, and again connect the eccentric with the rim of the dead wheel by inserting the eccentric link H into the other pair of lugs marked b. and d, using care in replacing the pins not to drive them so tight as to bind. Replace the set-screw in the outer side of the weight in exactly the same position as before. This reverses the governor and the engine is ready for service.

HOW TO KNOW STEEL FROM IRON

A drop of aqua fortis turns steel brown, and cast-iron black, while (wrought) iron is not affected.

AUTOMATIC GOVERNOR FOR SELF-CONTAINED ENGINES

SUCH AS DOUBLE CRANK DISC AND TWO PULLEY WHEELS, ONE EACH SIDE OF CRANK PIN

DESCRIPTION: In the following cut, B is the eccentric, hung on the hub M of the band wheel by a pin, which is made tapering to provide means for taking up the wear. The hole through the eccentric is much larger than the shaft and permits the center of the eccentric to shift across the shaft, thus varying its throw. Piece C is the eccentric arm, which transmits the shift motion to the eccentric through the steel bands, or ribbons, which are fastened by clamps to the eccentric arms and by screws to the eccentric. The arm is securely fastened to a rocker shaft, which passes through one of the arms of the band wheel, and to which the spring crossheads D, D are also attached. The spring crossheads carry the weight bars I, I, weights G, G, and the spring rods J, J. The springs K, K are held by the spring rods and are precisely alike, acting together as one. The weights, as well as the springs, are duplicated only to secure more perfect balance of the governor.

TO SET THE VALVE: Governor should set on

the shaft so that the center of the rocker shaft, to which the eccentric arm C is attached, is on the other side of the main shaft, directly opposite the crank pin, and keyed in that position. The steam valve controlled by the governor should be

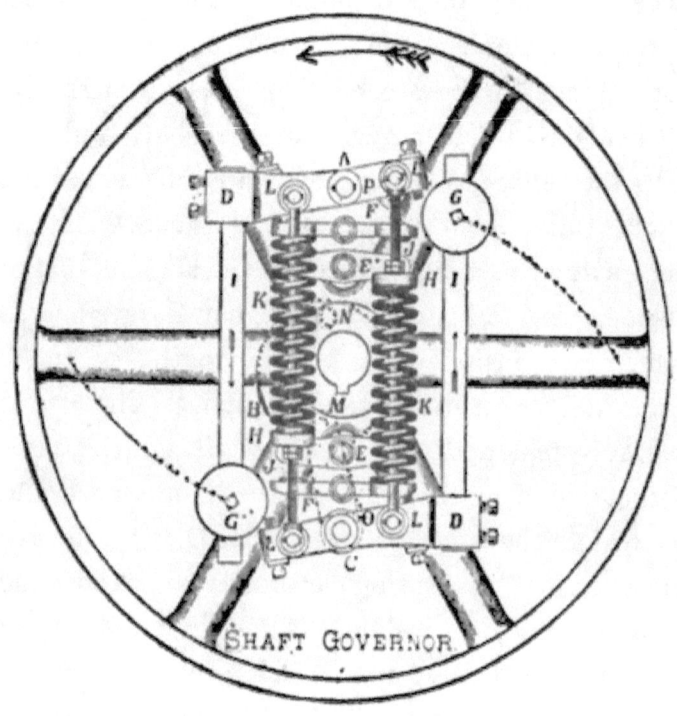

SHAFT GOVERNOR

adjusted to uncover the ports an equal distance on each end and should be set while the weights are out as far as the stops will allow them to go.

Place the engine on the "inner" center and allow the valve to just cover the steam port at the outer end of the cylinder so the outer end of the valve will be line and line with the outside edge of the outer port. Turn the crank pin in the direction it is to run to the "outer" center, and the inner end of the valve should correspond in like manner with the outside edge of the inner port. If it does not, equalize the difference by the nuts on the valve rod.

Roll the engine forward again to the "inner" center to be sure that the adjustment is right, and when this is accomplished the valve is set.

MANAGEMENT AND CARE should be the same as for side crank engine governor.

TO CHANGE SPEED: Same as for side crank engine governor. The speed may also be changed by altering the positions of the weights on the bars. Sliding them toward the spring crossheads increases the speed, and in the opposite direction decreases it.

Should a greater speed be desired than can be obtained from the springs and weights furnished with the governor, see instructions for speed changing on side crank engine governor.

To REVERSE governor for running over or under: Remove the eccentric strap, then take out the key from the governor band wheel, slip the wheel out to the end of the shaft, remove the taper pin on which the eccentric swings, move the eccentric to the other hole in the hub, replace the taper pin in the eccentric, using care that it is not screwed in so tight as to bind, change the weight bars to the other ends of the spring crossheads, reversing their positions, loosen the nuts that hold the tension on the springs, being careful to measure the springs before removing the nuts, so as to replace them in exactly the same positions.

After the springs are free from tension take out the small split pins that hold the ends of the spring rods in their places, remove both spring rods and turn them end for end, then replace the split pins.

Before tension is again put on the springs, move the weight bars back and forth to see that there is no binding in any of the working parts. Also see that the eccentric travels its entire throw across the shaft, and that both of the lips strike the top plate that is bolted to the face of the wheel hub, back of the eccentric. Should the eccentric strike on one point and not on the other, loosen the clamps that hold the steel bands, or ribbons, move the eccentric just far enough to enable both points to touch, then refasten the clamps.

Should the set-screws under the spring crosshead strike before the eccentric touches the stop, adjust them accordingly.

Place the required tension on the springs, slip the wheel back to its place, drive in the key, replace the eccentric strap and the engine is ready **to start.**

BALANCED SLIDE VALVE

Q.—What do the following cuts represent?

A.—The balanced slide valve of an automatic self-contained engine, protected from steam pressure by a hood.

Q.—What does the first cut represent?

A.—It is the valve and seat with hood detached.

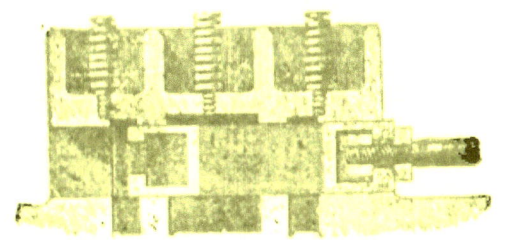

Q.—What does the second cut represent?

A.—A section through the valve and hood on the line A B, shown in the next cut.

Q.—What does the third cut represent?

A.—It is a perspective view of the valve and hood complete.

Q.—How is the balancing of valve accomplished?

A.—The exposed ends of valves, being of equal area, balance each other.

Q.—How is the pressure counterbalanced?

A.—By recesses of equal area with the ports under the hood and over the valve.

Q.—What friction is there to overcome?

A.—The only friction is the weight of a very light valve.

Q.—How is a worn valve refitted?

A.—By scraping.

Q.—What is provided for the case of excessive pressure or water in the cylinder?

A.—The hood is set loose on the seat, and readily yields.

Q.—What guides the hood to its correct seat?

A.—The springs and studs.

Q.—What is the port action of the valve?

A.—It is that of any plain slide valve.

Q.—What quickens the opening and closing of the ports?

A.—The recesses in the hood over the valve.

Q.—What advantage has this valve?

A.—All the advantages of a **perfectly balanced** slide valve.

CORLISS ENGINE

The Corliss valve gear is a detachable gear. There are four valves—two steam valves and two exhaust valves all connected to one center wrist plate. The wrist plate pin is connected to rocker arm by reach rod, and from there to eccentric by another rod.

1. Fly ball gov.
2. Gag pot.
3. Gov. stand.
3A. Bevel gear case.
4, 4. Gov. rods.
5, 5. Steam valves.
6, 6. Exhaust valves.
7. Wrist and rocker connecting rod.
8, 8. Dash pots and rods.
9. Eccentric rod.
10. Governor belt and pulleys.
11. Eccentric and strap.
12. Rocker arm.
13. Wrist plate.
14. Cylinder Bracket.
15, 15. Steam valve (adjusting) rods.
16, 16. Exhaust valve rods.

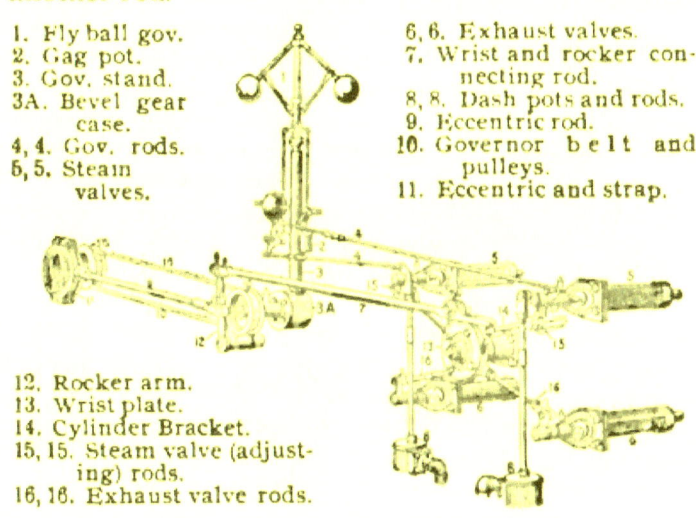

TYPICAL CORLISS VALVE GEAR

To SET VALVES, take off the back caps or back heads of all four valve chambers. Guide lines will be found on the ends of the valves and chambers, as follows: On the steam valves, lines indicating the working edges of the steam ports; on the exhaust valves and ports, guide lines for

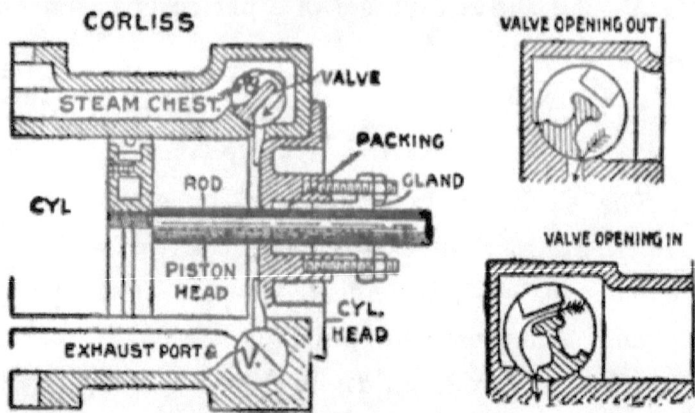

the purpose of setting them. As stated before, the wrist plate is centrally located between the four valve chambers on the valve gear side of the

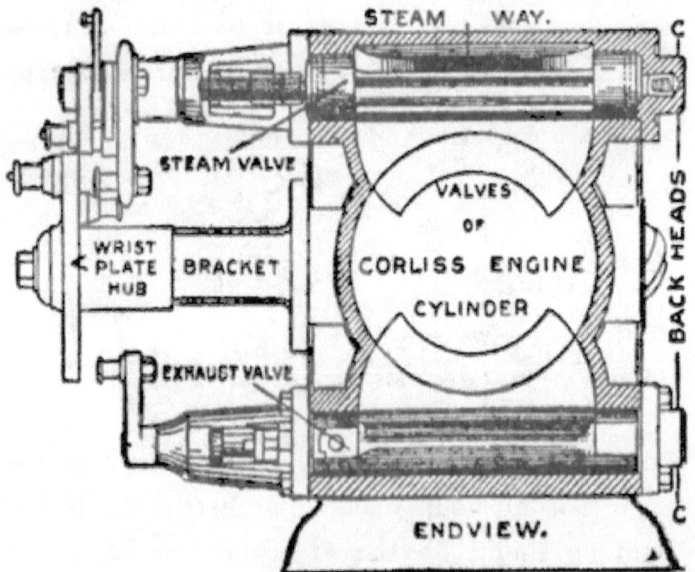

cylinder. A well-defined line will be found on the bracket which is bolted to the cylinder, and three lines on the hub of wrist plate, which, when they correspond with the single line on the bracket, show central position of the wrist plate, and the extremes of its throw or travel both ways.

To ADJUST THE VALVE first unhook the reach or carrier rod connecting the wrist plate with rocker arm, then hold the wrist plate in its central position.

The connecting rods between steam and exhaust valves and wrist plate are made with right and left hand screw threads on their opposite ends, and provided with jamb nuts, so that by slacking

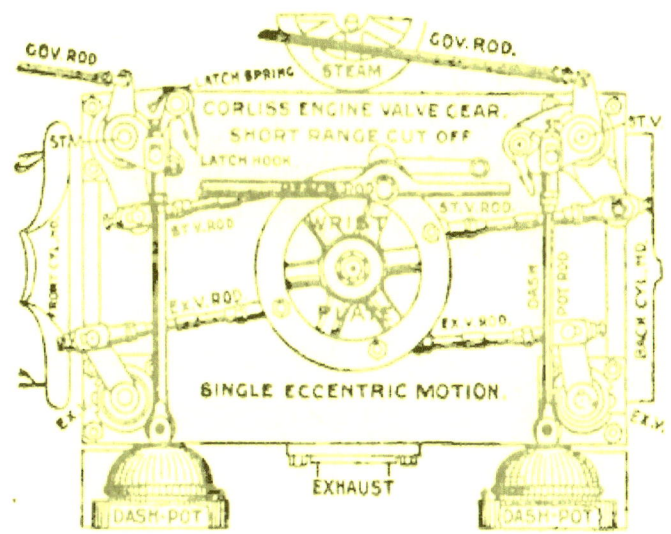

the jamb nuts and turning the rods they can be lengthened or shortened as desired. By means of this adjustment set the steam valves so that they will have ¼ inch lap for 10 inch diameter of cylinder, and ½ inch lap for 32 inch diameter of cylinder, and for intermediate diameters in proportion.

For the Exhaust, set them with 1-16 inch lap for 10 inch bore, and ⅛ inch lap for 32 inch bore on non-condensing engines, and nearly double this amount on condensing engines for good results. Lap on the steam and exhaust valves will be

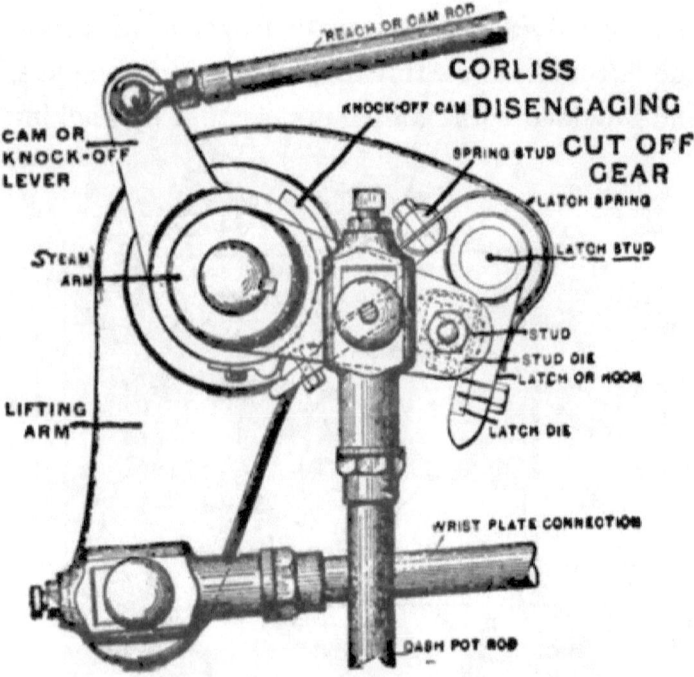

shown by the lines on the valves being nearer the center of the cylinder than the lines on the valve chambers.

Having made this adjustment of valves, the rods connecting the steam valve arm with the dash pot should be adjusted by turning the wrist plate to its extremes of travel and adjusting the rod of each valve so that when it is down as far as it will go the square steel block or stud die on the valve arm will just clear the latch die on the latch hook.

If the rod is left too long the steam valve stem would likely be bent or broken; if too short, the hook will not engage, and, consequently, the valve will not open.

Having adjusted the valves as stated, hook the engine in, and, with the eccentric loose on shaft, turn it over and adjust the eccentric rod so that the wrist plate will have the correct extremes of travel, as marked on the wrist plate hub.

If marks on wrist plate do not agree at each full throw with bracket marks, disconnect strap from eccentric rod and adjust the screw on stub end, as required, until marks do agree, both forward and backward; then place the crank on dead center and turn the eccentric in direction engine is to run, until an opening of 1-32 or 1-16 is shown at steam valve, then throw crank pin on other dead center to secure the desired lead in opposite motion. If lead is not the same, adjust by lengthening or shortening the connecting rods between the eccentric and wrist plate as the case may be.

To ADJUST THE RODS connecting the cut-off or tripping cams with the governor, have the governor at rest and the wrist plate at one extreme of its travel. Then adjust the rod connecting with the cut-off cam on the opposite steam valve,

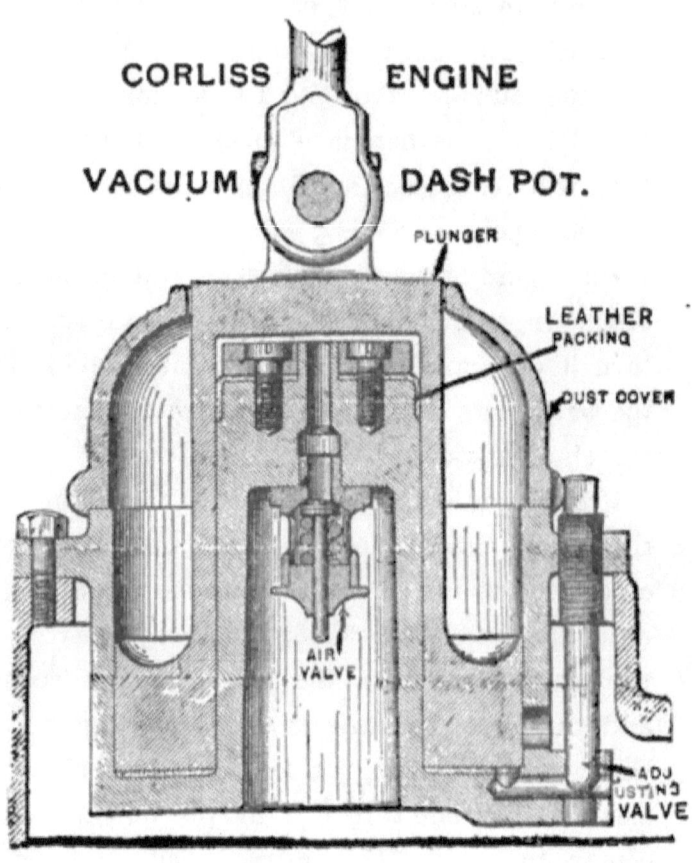

so that the cam will clear the steel or latch die on the tail of the hook about 1-32 of an inch. Turn the wrist plate to the opposite extreme of travel and adjust the cams for the other valves in the same manner.

CORLISS ENGINE

To EQUALIZE THE CUT-OFF and test its correctness, hook the engine in and block the governor up about halfway in the slot, which will bring it to its average position when running. Then turn the disc slowly in the direction which it is to run, and note the distance the crosshead has traveled from its extreme position at dead center when the cut-off cam trips or detaches the steam valve. Continue to turn the disc beyond the other dead center and note the distance of crosshead's extreme travel when valve drops. If distance is the same the cut-off is equal; if not, adjust either one or the other of the rods until the distance is the same.

REVIEW

Q.—Will the cut-off mechanism unhook when governor is down?

A.—No; it keeps the valve hooked up full stroke.

Q.—Will the latch die hook on stud die of valve when dash pot rod is too short?

A.—No.

Q.—How long should the rod be?

A.—It should be long enough so when the plunger is at the bottom of dash pot the latch would hook over the latch stud (steel block) and the stud lie clear of the latch (hook).

Q.—What prevents the dash pot rod from breaking or bending?

A.—The cushion of the plunger on air in the dash pot.

Q.—Have the plungers any packing to make them close fit?

A.—Yes; some have leather packing, others have piston rings.

Q.—How is the air regulated in the dash pot?

A.—By means of an air valve in the air opening by turning a screw in the escape hole.

Q.—What keeps the governor in regulation so it will not allow the engine to run away or be oversensitive?

A.—A small oil reservoir on engine frame below governor, known as the gag pot.

Q.—What kind of oil is generally used in the gag pot?

A.—Kerosene oil.

Q.—How would you give the governor more freedom of motion?

A.—By removing one or more of the small screws in the piston plunger of gag pot.

Q.—How would you warm the cylinder of a Corliss engine before starting?

A.—By first blowing all the condensed water out of the steam pipe by means of the drip valve provided on the steam valve elbow or globe; then open the steam valve a little to allow the valves and cylinder to become warm. Unhook rocker reach rod, and work valves with wrist plate by nand with lever. The cylinder soon becomes warm and all water is expelled into the exhaust pipe, the exhaust drain cock having been left open to allow the condensed water to escape.

Q.—Would you then start the engine up lively?

A.—No. Let engine move slowly until satisfied all is right, then open throttle gradually until wide open.

Q.—Suppose the governor belt connection broke, would the engine run away?

A.—No; the trips or safeties would slip in between the latches and dies and prevent valves from opening or latches hooking on to latch dies.

Q.—Suppose the governor of a Corliss engine would allow the speed to fluctuate from one extreme to the other, where would you look for the trouble?

A.—The oil gag pot in connection with the governor will very probably be found to be empty, on inspection.

Q.—In the majority of cases where the governor gives trouble, what would you lay it to?

A.—Not getting proper oiling, being dirty, oil holes plugged and not good enough connection to the main shaft.

Q.—About how many oil holes has a Corliss governor, all told?

A.—From 10 to 13.

LINK MOTION AND VALVE SETTING

In the position of the link shown in the cut, the valve has its shortest travel. The further removed the link is from the position shown, either upward or down, the longer is the valve's travel.

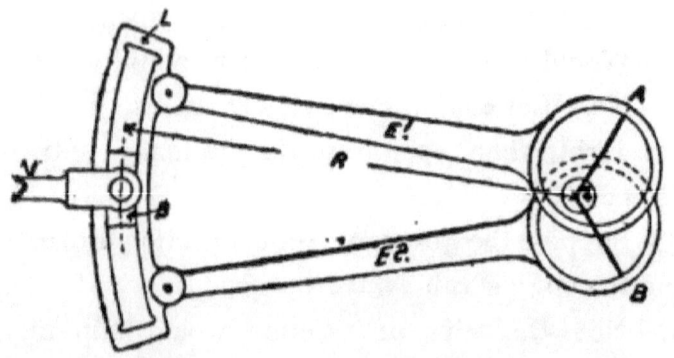

V. Valve Stem.
B. Link Block.
L. Link.
E1 E2. Link Blades.
R. Radius.
S. Shaft.
a. Heaviest side of backward over eccentric.
b. Heaviest side of forward under eccentric

FOR VALVE SETTING in a stationary engine, 12x24 inches, place the valve central over the ports, the rocker arm plumb, the heavy side of the eccentric plumb over the shaft, and the crank pin at dead center. See that the eccentric blades are connected with the link and are in full gear, forward or backward. This will make the extreme travel of the valve equal to the throw of the eccentrics. If a lead of 1-16 of an inch is desired, move the eccentric in the direction in which you want the engine to run, until your valve has the desired lead. Then fasten the eccentric, throw the crank pin on the opposite dead center and if the lead on the opposite port is then the same, the valve is set. If it is not, make adjustments.

The above covers one motion. **For the opposite** motion reverse the link and go through the same operation for the other eccentric.

For convenience an engineer should tram his valve-stem, so as to know the opening point either way without removing the valve chest cover.

Q.—Why is a link placed on an engine?

A.—Because it is the most convenient means for reversing an engine. It is almost a necessity where a valve has much steam lap, or where quick reversing is required.

Q.—Which way does the engine run when the link is fully down on the block?

A.—It would run under, toward the cylinder.

Q.—How would you reverse the engine in that case?

A.—By pushing it full up.

Q.—How can a single-eccentric engine be made to be easily reversible?

A.—By substituting a rocker arm with another rocker pin above the center of rock shaft. (See page 101.)

AREA OF STEAMPORT.

The area of the steamport may be justly considered as the **basis** from which all other dimensions are derived in conformity with known laws.

It makes a difference, whether the port is simply to admit the steam to the cylinder, or whether it is also to serve as exit or exhaust.

In the latter case a small quantity of steam forces its way out with a **constantly diminishing** pressure and, therefore, the exhaust port must be larger than the steam port.

Where the same port serves for both purposes, it must have the proper area for the exhaust, and is opened only partly for the admission of steam, which enters from the boiler with a practically constant velocity.

NUMBER OF CRANK REVOLUTIONS FOR GIVEN STROKE AND PISTON SPEED.

STROKE.	PISTON SPEED (feet per minute.)									
	200	210	220	225	230	240	250	270	300	350
1 ft. 6 in.	67	70	73	75	76	80	83	90	100	116
1 " 8 "	60	63	66	68	70	72	75	81	90	105
1 " 10 "	55	57	60	61	63	66	68	74	82	96
2 " 6 "	40	42	44	45	46	48	50	54	60	70
3 " 0 "	33	35	36	37	38	40	42	45	50	58
4 " 0 "	25	26	27	28	29	30	31	34	38	44
5 " 0 "	20	21	22	22	23	24	25	27	30	35

HORSE POWER

Coal furnishes heat; heat converts water into steam; the steam drives the piston, the piston motion is converted into rotary motion by the connecting rod and crank pin. The rotary motion is utilized for work. The amount of work that an engine can do is expressed in "horse-powers," the unit being determined as follows:

The usual traveling gait of a horse hitched to a light sulky is about 5 miles an hour, or 440 feet per minute. If a spring scale be attached to the singletree we may note the amount of power the horse is exerting. Assuming this to be 75 lbs. and the speed 440 feet, multiplied by 75 lbs. equals 33,000 foot lbs., which represents a horse-power. In applying this to a steam engine we first find the area of the face of piston head, multiply the answer by piston speed in feet per minute, and divide by 33,000; the answer will be the indicated horse-power. For the actual or effectual horse-power take 2-3 of the quotient.

Example: Engine cylinder 12x24, speed 100 revolutions per minute, steam 80 lbs., area of piston 113 square inches. Multiply 113 by 80, equals 9,040 lbs. pressure on piston face, by 400 feet piston travel per minute, equals 3,616,000, divided by 33,000 equals 109 N. H. P. full opening of valve, deduct 1-3 for cut-off, equals 72 2-3 actual h. p. For short cut-off one-half. The reduction is made for average pressure, condensation, friction, etc., and will be found quite correct in practice.

Quick rules, such as are generally given for finding the H. P. of cylinders, waterfalls, waterwheels, etc., are useless.

One "watt" is the 1-746 part of one horse power. One thousand watts or a "kilowatt" equals one and one-third horse power. The watt is the practical unit of electrical activity or power; it is the rate of working in a circuit when E. M. F. is one volt and the current one ampere. (See electricity.)

The best engines and boilers develop a horse power per hour by the consumption of 2 lbs. of coal. But this is better than the average, and 3 lbs. is more common.

Q.—How much heating surface is required to develop one horse power?

A.—It varies with the purpose of the plant.

Steam for heating, etc............15 sq. ft. heating surface
For plain throttle engine............15   "   "   "
For simple Corliss engine..... . ..12   "   "   "
For compound Corliss condensing..10 "   "   "

Q.—How many horse power will a boiler furnish

for a plain slide valve engine, boiler having 1,500 square feet heating surface?

A.—One hundred H. P.

Q.—How much for simple Corliss engine, same boiler?

A.—One hundred and twenty-five H. P.

Q.—For compound engine?

A.—One hundred and fifty H. P.

Q.—Which would you consider the best basis in comparing boilers?

A.—Their evaporative efficiency.

Q.—Give consumption of steam per indicated horse power per hour for various engines?

A.—Plain slide valve engine........60 to 70 lbs.
High speed automatic engine. 30 to 50 "
Simple Corliss engine......... 25 to 35 "
Compound Corliss engine. ....15 to 20 "
Triple expansion engine........13 to 17 "

An engine of the proper size and in good condition will yield one H. P. at the lowest consumption.

Q.—How would you determine the proper size or evaporating capacity of a boiler to supply steam for a given purpose?

A.—It is necessary to consider the number of pounds of dry steam actually required per hour at stated pressure.

Q.—What is the standard horse power rating for any steam boiler for common slide valve and Corliss engine?

A.—For plain slide valve engine the evaporation

is 62½ lbs., or one cubic foot of water per hour per horse power, and for the Corliss 31¼ lbs., or ½ cubic foot of water per hour per horse power.

Q.—How would you figure the horse power for a steam boiler of any size, if you wish to run a 30-horse power engine for 1 hour, carrying 60 lbs. of steam pressure?

A.—First multiply the pressure to be carried by time in minutes, 60, and divide by 30, the amount of water in pounds per horse power evaporated per hour.

Q.—How would you proceed to find the horse power of a compound condensing engine?

A.—The H. P. of a compound condensing engine, of necessity, cannot amount to any more than the aggregate of the two powers produced in the two cylinders. Therefore the power developed in each cylinder must be calculated separately and the two results added. (See indicator, p. 154.)

Q—What is good working vacuum for a steam engine?

A.—From 22 inches upward, when the barometer stands at 30 inches.

Q.—Suppose the M. E. P. upon the piston is 40 lbs. per sq. in., and the vacuum gauge stands at 22 inches (barometer at 30 in.), what would be the total on one side of the piston?

A.—40 lbs. per sq. inch, the M. E. P.

Q.—How is the horse power found of a non-condensing compound engine?

A.—By first finding the average area of both cylinders. This is done by finding the area of the high and the low pressure cylinders separately; then add them both together and divide by two.

Q.—How is the average mean effective pressure found?

A.—By finding each cylinder's mean effective pressure, then adding the two together and dividing by two.

Q.—After this is done, how would you proceed to find the gross H. P.?

A.—Multiply the average area by the average mean effective pressure (see page 159), then by the piston travel per minute, and divide by 33,000. Answer will be H. P.

Q.—How do you increase the power of an engine.

A.—By increasing its speed. It is done:

First, by increasing the boiler pressure, and, thereby, the M. E. P. on the piston.

Second, by increasing the boiler pressure and decreasing the outside lap of valve so as to cut off late. This method is not advisable.

Third, by increasing the leverage of the main shaft pulley by decreasing its diameter.

Fourth, by increasing the speed of line shaft or the diameter of the driver pulley on line shaft, or by decreasing the diameter of pulley on main shaft.

Q.—Give the rule for finding the horse power of a belt's transmission; also example?

A.—Multiply the width of the belt in feet by the number of hundred feet the belt has traveled in one minute. Example: Belt 2 feet wide running 150 feet per minute—2 multiplied by 150 equals 300 h. p.

Belting horse power of a belt equals velocity in feet per minute, multiplied by the width. One inch in length of single belt moving at 1,000 feet per minute per 1 inch width equals 1 h. p. For double belts of great length over large pulleys allow about 500 feet per minute per 1 inch of width per horse power. Power should be communicated through the lower running side of a belt, the upper side to carry the slack.

The average breaking weight of a belt 3-16x1 inch, single leather, is 530 lbs.; three-ply rubber belt, 600 lbs. The strength of a belt increases directly as to its width. The allowance for safety for rubber belts is $\frac{1}{8}$ and for leather belts 1-16 (breaking weight) in lacing.

Q.—Can you give a short rule for finding the H. P. of tubular boilers, and are such rules of value?

A.—They are not. A short rule to find the horse power of a tubular boiler is: Multiply the square of diameter in feet by length and divide by constant .4. For flue boiler multiply diameter of shell in feet by length and divide by .4, or multiply area of grate surface in square feet by 1½. The answer gives the horse power.

Table Giving Horse Power of Boilers of the Usual Sizes.

Diameter Shell. Inches.	Length Shell. Feet.	Number Tubes.	Length Tubes. Feet.	Diameter Tubes. Inches	Heating Surface. Square feet.	Horse Power 80 lbs. Pressure.
72	18	70	18	4	1502	100
72	16	90	16	3½	1472	98
72	16	112	16	3	1496	99
72	15	112	15	3	1400	93
60	18	65	18	3½	1200	80
60	17	65	17	3½	1148	76
60	16	65	16	3½	1075	72
60	16	80	16	3	1088	72
54	18	50	18	3½	951	63
54	17	50	17	3½	900	60
54	16	50	16	3½	795	53
54	16	60	16	3	832	55
48	16	40	16	3½	683	46
48	16	49	16	3	684	46
48	15	49	15	3	642	43
48	14	49	14	3	600	40
42	15	38	15	3	508	34
42	14	38	14	3	476	32
42	13	38	13	3	441	30
42	12	38	12	3	408	27
42	11	45	11	2½	390	26
42	10	45	10	2½	355	24
42	9	45	9	2½	320	22

THE INDICATOR

Q.—What is an indicator?

A.—It is an instrument which records the variations of pressure during the length of one stroke.

Q.—Can you describe the instrument?

A.—A small cylinder of exactly ½ sq. inch inside diameter is connected to both ends of the steam cylinder, but steam is admitted from one end at a time only. In the small cylinder moves a piston, whose crosshead works a pair of light levers, the free end of which holds a pencil, which marks its path of motion on a paper clamped on a revolving drum.

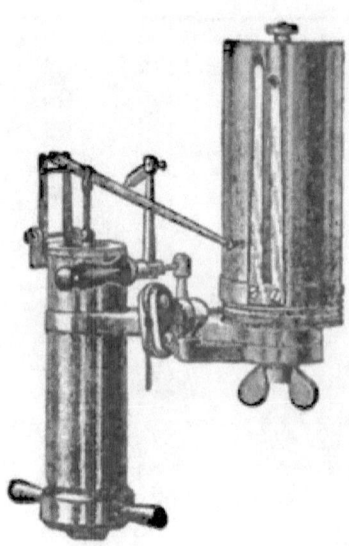

Q.—How is the stroke of the instrument's piston regulated?

A.—By a spring of known tension. A set of such springs, each marked with the pressure for which it is intended, accompanies each instrument, as follows:

For pressure up to 21 lbs. per sq. inch use 15 lb. spring.
"   "   "   " 38   "   "   "   "   " 20   "   "
"   "   "   " 94   "   "   "   "   " 30   "   "
"   "   "   " 143   "   "   "   "   " 50   "   "

Q.—What makes the indicator card revolve?

A.—A carefully adjusted cord, indirectly connected to the crosshead of the engine.

Q.—How do the **pencil tracings** on the paper convey information?

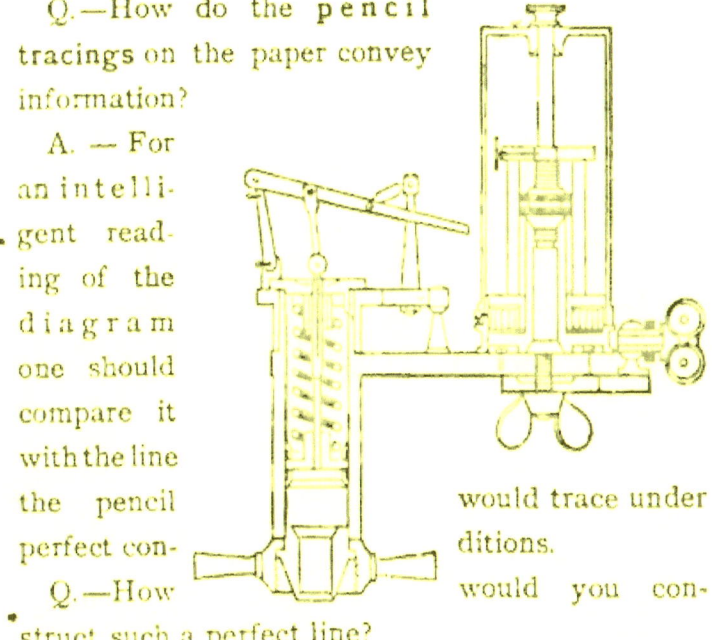

A. — For an intelligent reading of the diagram one should compare it with the line the pencil would trace under perfect conditions.

Q.—How would you construct such a perfect line?

A.—Assuming an engine to have a 32″ stroke, 60 lbs. steam (gauge pressure), vacuum 12 lbs., cutting off at 8″, exhaust release 2″ from end of stroke, and compression (exhaust closure) 5″ from completion of stroke,—I should lay off these figures on a "cross-section" sheet. (Fig. 1 shows the cross-section in the margin only, to give a clearer cut. Only the principal lines are drawn all across.)

Mark off 32 spaces, each to represent 1″ of stroke, horizontally, calling the starting point at the left 0 and the end at the right 32. Mark off vertically 25 spaces, each to represent 3 lbs. of pressure. The upper limit of the fifth space (starting from the 0 point first mentioned) will

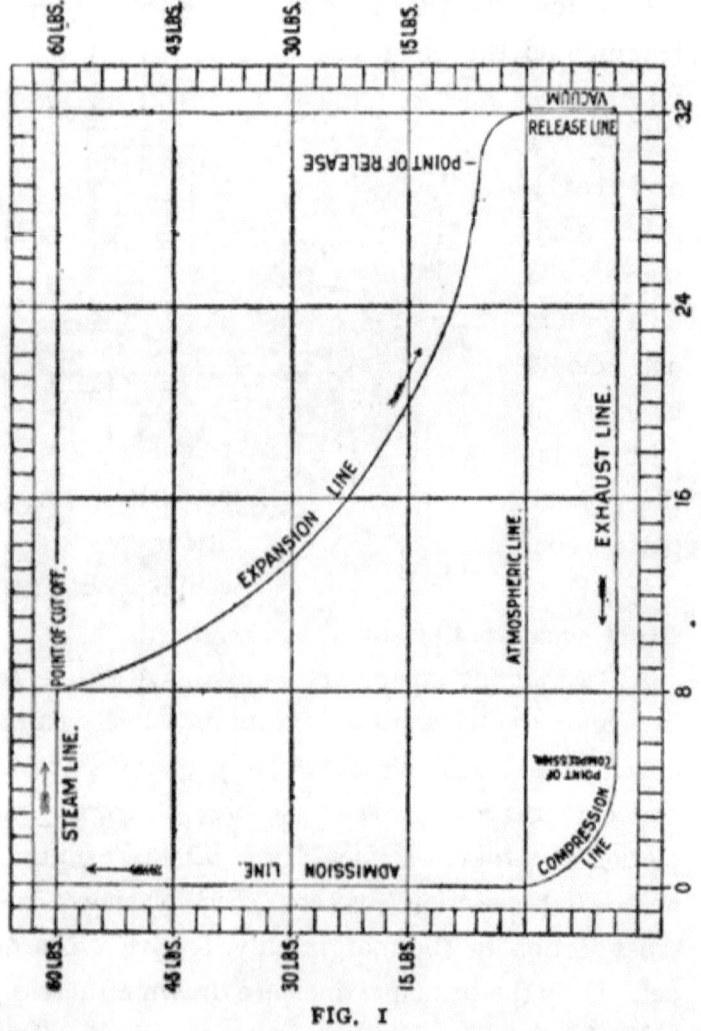

FIG. 1

then be atmospheric line (3×5=15), 5 spaces further will be the 15-lb. steam pressure line, and so on, the last line representing the 60-lb. pressure.

Steam enters in our case during ¼ of the stroke, therefore, we make the **"steam line"** 8 spaces long on the 60-lb. line, starting from the 0 vertical. At the end of the steam line the supply is **cut off, expansion** begins and pressure is reduced, first rapidly, then more and more slowly (in an inverse ratio to the volume; at point 16 the volume has been doubled and the pressure halved). This gives an evenly curved line down to the intersection of the vertical 30 (2° from end of stroke) with the horizontal indicating 6 lbs. of pressure (2 spaces above the **atmospheric line**). At this point (**point of release**) the release (exhaust) valve should open, and the pressure sink again rapidly down to nothing (atmospheric line) and below, represented by a short, sharp curve between the verticals 30 and 32 and a straight line along the vertical 32, 4 spaces to the exhaust line (**release line,** or **vacuum line**).

The engine piston travels then 27 inches in the opposite direction, while the exhaust valve keeps open (**exhaust line**) to the **"point of compression"** where the not exhausted steam begins to be compressed until during the last 5 inches of the piston's travel the pressure is gradually brought up to atmospheric pressure. The "compression line" representing this is an upward curve ending at the intersection of the atmospheric line and the zero vertical. At this point **steam is admitted**, raising the pressure *instantly* to 60 lbs., shown on the diagram by a straight line along the vertical zero up to the 60-lb. horizontal. This completes the diagram.

Q.—Does the diagram taken on engines deviate much from this model just described? and if so, why?

A. — There are great deviations, caused by leaks, wrong lead, late valves, light load with heavy compression, etc.

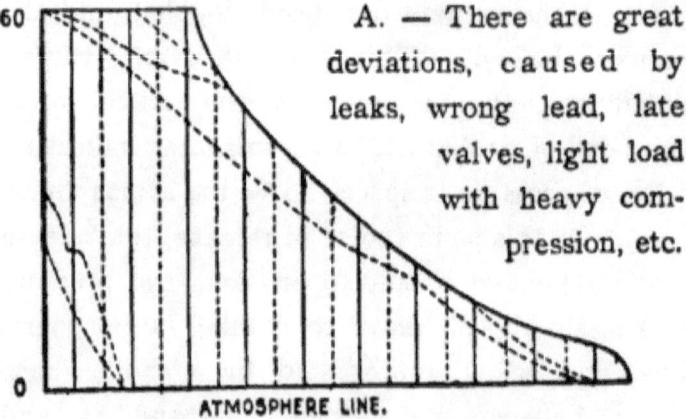

FIG. 2

(The broken lines in the cut give examples of diagrams indicating imperfections in the cylinder or valves, while the outline is a nearly perfect **specimen of** diagram above atmospheric line)

Q.—What is the purpose of an indicator card?

A.—It serves as a guide in setting the valves, as a help (in connection with a feed-water test for steam-consumption) in determining the economy with which an engine works, and especially for finding the MEAN EFFECTIVE PRESSURE of an engine.

Q.—How do you figure the M. E. P. from an **indicator card?**

THE INDICATOR

A.—Divide the extreme length of the diagram in 10 equal spaces vertically by 9 dotted lines (Fig. 2), and divide each space into vertical halves by full lines. The length of these 10 vertical lines (ordinates) inside the diagram indicates the M. E. P. for each space. Add these 10 lengths together, and divide their sum in inches by 10. Multiply the quotient by the scale of the spring used in the indicator, and the product will be the M. E. P. throughout the stroke.

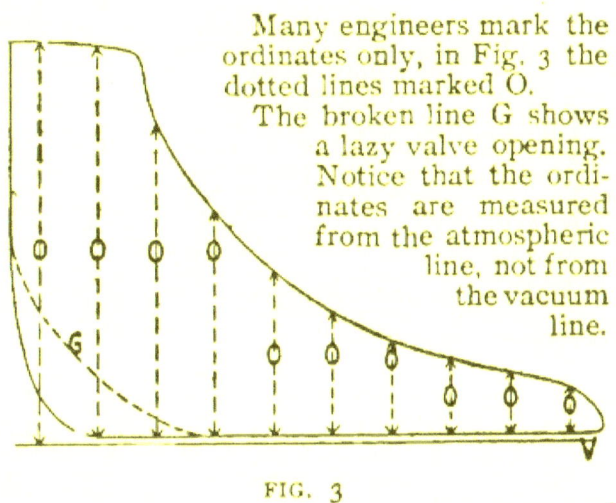

Many engineers mark the ordinates only, in Fig. 3 the dotted lines marked O.

The broken line G shows a lazy valve opening. Notice that the ordinates are measured from the atmospheric line, not from the vacuum line.

FIG. 3

Q.—Why do you multiply by the scale of the spring?

A.—Because one inch of height in the diagram represents the amount of pressure indicated on the spring, two inches the double, etc.

Q.—Is there a quick way of figuring the M. E. P.?

A.—Yes. Make a rough sketch of a diagram and divide the length of it into 10 equal spaces; allow for the first four spaces (to cut-off) the full pressure as per gauge, say 100 lbs. each, divide their sum, 400, by 5, the number of the next space,

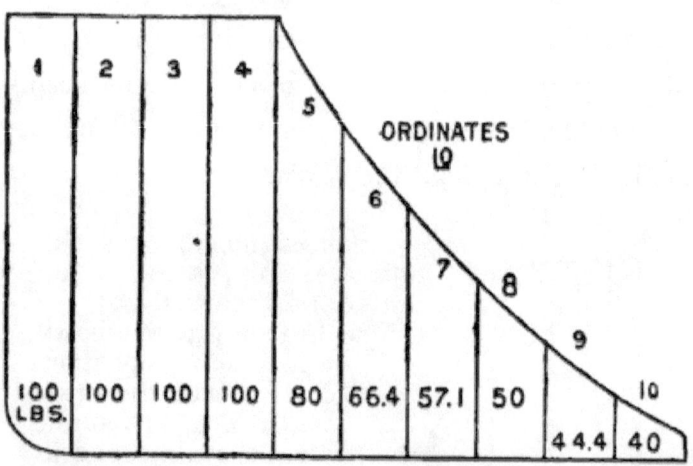

allowing the quotient, 80, for the fifth space; then divide the same sum, 400, by 6 and allow this quotient for the sixth space, and so on, to the last space. Add all these figures together, and divide by 10 and proceed as above. (See cut.)

Q.—Is this a very accurate way?

A.—No, but it will answer for a rough figuring under ordinary circumstances.

Q.—How is the PANTOGRAPH used in connection with the indicator?

A.—One end of it (C) is fixed to the crosshead, the other (D) is made stationary by a fixed stake

THE INDICATOR   161

placed in line with the crosshead socket at midstroke. A peg (F) is placed in one of the holes on the adjustable strip G, so as to be on the line between

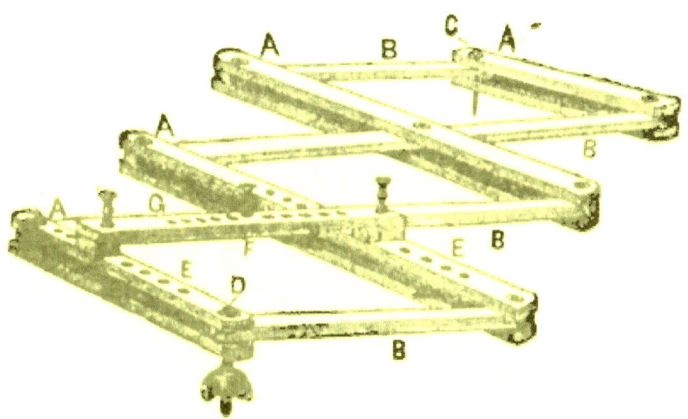

the two points C and D, and at such distance from C that the cord connecting it with the indicator drum will be parallel with the guides.

The peg will not move as fast as the piston head, but it moves at exactly the same ratio, giving an accurate diagram. (See cuts.)

INDICATOR EXAMINATION

Q.—Can any one use the indicator intelligently?

A.—No, only one who has had experience with engines, who possesses power of observation, and who is familiar with measurements and calculations.

Q.—What length can a diagram be?

A.—As long as the circumference of the drum.

Q.—What do the length and the height of the diagram represent?

A.—The length represents the length of the stroke. The diagram is one inch high for every 15 or 20, etc., lbs. pressure, according to the spring scale.

Q.—If the 30 lbs. spring is used and the diagram is $2\frac{1}{8}$ inches high, what does this indicate?

A.—It would indicate that the greatest pressure during the stroke (steam line) was $2\frac{1}{8}$ times 30 lbs.

Q.—Would the 15 lbs. spring answer in this case?

A.—No. The pressure would be $63\frac{3}{8}$ lbs., while the 15 lbs. spring is not able to act under a pressure of more than 21 lbs.

Q.—Explain the steam line?

A.—It runs from the place of admission to beginning of cut-off.

Q.—Where is the exhaust line?

A.—It begins at the point of exhaust or release.

Q.—Which is the expansion line?

A.—It is the curved line between the cut-off and the point of exhaust.

Q.—What is the vacuum line?

A.—A straight line, laid out by measuring down from the atmospheric line a distance equal to pressure of atmosphere, as shown by the barometer.

Q.—Does this line indicate a real vacuum?

A.—No, it indicates a reduction of the atmospheric pressure.

Q.—What points does an indicator card show?

A.—The high and low pressure, cut-off and lead, exhaust point and atmospheric point.

Q.—How does the steam line show on a card when steam is wire drawn?

A.—It falls as the piston advances.

Q.—What is meant by wire drawing?

A.—It is reducing the pressure by choking.

Q.—When should the atmospheric line be taken on the card by indicator?

A.—Immediately after the card has been taken.

Q.—Why?

A.—Because the spring in cooling will change the position of the pencil point.

Q.—How is the atmospheric line drawn?

A.—By holding the pencil lightly against the card, taking care not to get it out of its true position, and then revolving the drum with the hand.

Q.—What is an **ordinate?**

A.—It is the length of a line showing the height of a point above a level or line.

Q.—How many ordinates are marked on a diagram?

A.—You may mark any number. The larger the number the more accurate will be the result. Ten is the number usually taken for ordinary purposes.

Q.—Where do you place the ordinates?

A.—In the middle of the ten equal spaces into which I divide the diagram.

Q.—What does each single ordinate show?

A.—It indicates the mean effective pressure during the time in which the pencil passes across the space, in the middle of which the ordinate lies.

Q.—What is meant by "mean"?

A.—If the pressure was 60 lbs. on entering a space and ran down evenly to 55 lbs. on leaving the space, 57½ lbs. would be the half-way between the two, or the "mean," which may be safely called the pressure all across the space.

Q.—After the card has been properly laid out, what should be done?

A.—Measure the combined length of all the ordinates, divide their sum by the number of ordinates, and multiply the quotient by the figure on the spring scale in use.

Q.—How do you "add" the ordinates?

A.—By laying them off in one continuous straight line on a cardboard, then measuring their total length in inches.

Q.—Suppose the total length of the ten ordinates were 8 inches and the spring used was a 50 scale, how would you proceed?

A.—The eight inches and 10 ordinates equals eight-tenths, or .8, multiplied by 50 scale equals 40 pound mean effective pressure in the cylinder of engine.

Q.—Does the indicator allow for all friction, etc.?

A.—Yes.

Q.—How are the ordinates measured, when the expansion line drops below the atmospheric line?

A.—The sum of the lengths below the atmospheric line is subtracted from the sum of those above. Otherwise the calculation is the same.

Q.—What does the scale number on an indicator card mean?

A.—It indicates what pressure will lift the pencil point one inch. (See cut, page 156.)

Q.—What is meant by *mean effective pressure?*

A.—It means the average pressure of steam in the cylinder during one stroke, minus the back pressure.

THE COMPRESSED NON-VIBRATING AIR OR STEAM ENGINE

CAN BE USED TO PROPEL HORSELESS CARRIAGES, YACHTS, MOTOR WAGONS, ETC.

The growing demand for a small engine, more particularly for traction and marine work, simple in construction and operation, economical, non-vibrating, light weight, yet strong and compact, and reversible by simply shifting the valves, has resulted in the perfection of an engine far in advance of anything heretofore made for the purpose.

One of the more essential requirements for the above purposes is: Sufficient strength to start a given load from a standstill, which must be greater than the force necessary to overcome ordinary obstacles when in motion.

The "compressed air engine" develops extreme power for its weight and the space it occupies. It also dispenses with all vibration, which heretofore has been the great trouble in locomotion of horseless carriages, etc.

DESCRIPTION: The cylinder, as seen in following cut, has 2 ports, 3 pistons, 1 stuffing box, 1 crosshead and slide for same and connecting rods to a double crank arm shaft. No cylinder heads. This construction makes no difference in the

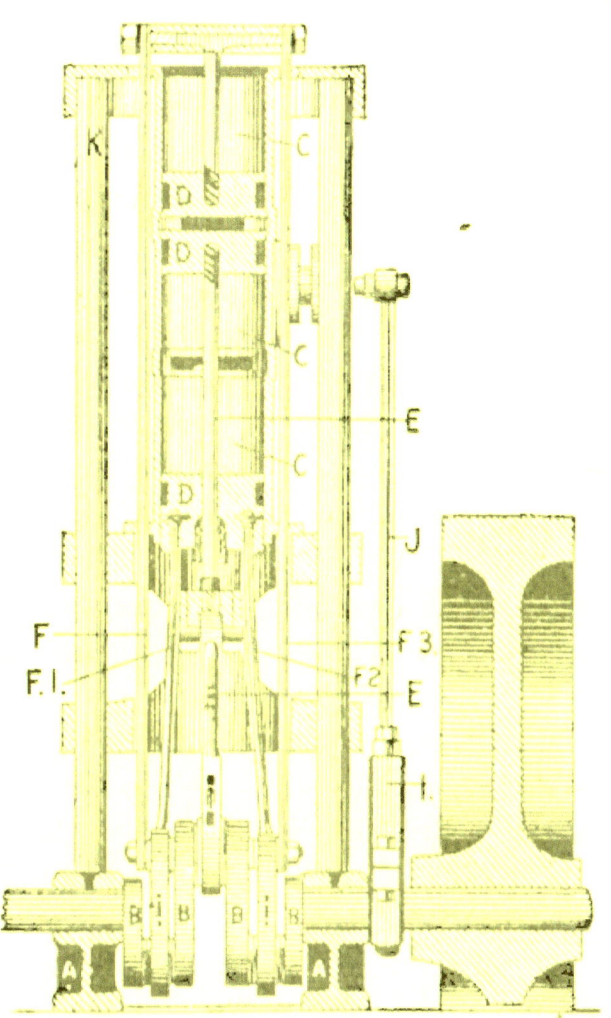

THE COMPRESSED NON-VIBRATING AIR OR STEAM ENGINE

A A, Bedplate; B B, Double Cranks and Shafts; C C, Cylinder; D D, 3 pistons; E E, Center Piston Rod and Connecting Rod; F F1 F2 F3, Four Connecting Rods; I, Eccentric; J, Eccentric Rod; K, Engine Tube Frame. The steam chest does not show, being behind the cylinder.

action, but it does in the results obtained. The two outside pistons are fastened together so as to move in the same direction, and are connected to one side of the crank shaft. The center piston, acting between these two, is connected centrally between two cranks and always moves in the opposite direction from the two outside pistons. It travels between the two ports, first meeting the lower piston at lower port, then reversing and meeting the upper piston at the upper port, the two outside pistons always traveling outside of the two ports, thus answering the purpose of the two cylinder heads, which are simply converted into movable heads or pistons.

In a starting pull, the same charge of steam or air acting on both pistons at the same time gives twice the power at starting and a double expansion for the balance of the stroke. The ports and exhaust are the same in construction and operation as standard makes of engines on the market.

MENDING A BAND SAW

Bevel both ends of the saw the length of two teeth. Fasten the saw in brazing-clamps, with the backs against the shoulders. Wet the joint with solder fluid (or with a lump of borax rubbed into a creamy paste with a teaspoonful of water on a slate). Put a piece of silver solder of the shape of joint in the joint, and clamp with tongs heated to a light red heat. As soon as the solder fuses, blacken the tongs with water (taking care not to get any water on the saw), release tongs and smooth the joint by hammering and draw-filing.

MISCELLANEOUS
QUESTIONS AND ANSWERS
ON THE ENGINE

TRAVEL OF CRANK PIN AND CROSSHEAD

Q.—Do crank pin and crosshead travel at an even gait during one revolution of the disc?

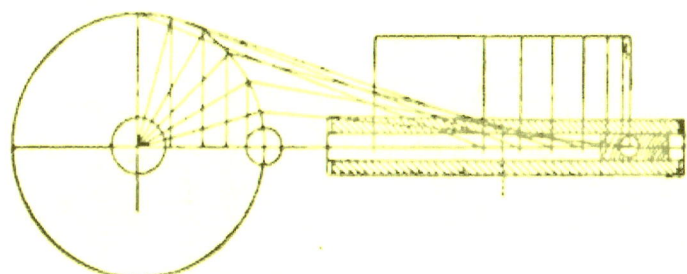

A.—They do not. The diagram shows (1) that the crosshead travels only a very short distance while the crank pin moves 15 degrees (⅙ of 90°, a quarter revolution) upward from the dead center; (2) that the distance traveled by the crosshead increases in each of the 5 following spaces, each of them corresponding to the crank pin's travel of 15 degrees (6x15° = 90°); (3) that the crosshead has traveled more than one-half of its stroke (the half-way point is indicated by a dotted line in the cut), when the crank pin has traveled 90°, or ¼ revolution.

170   MISCELLANEOUS

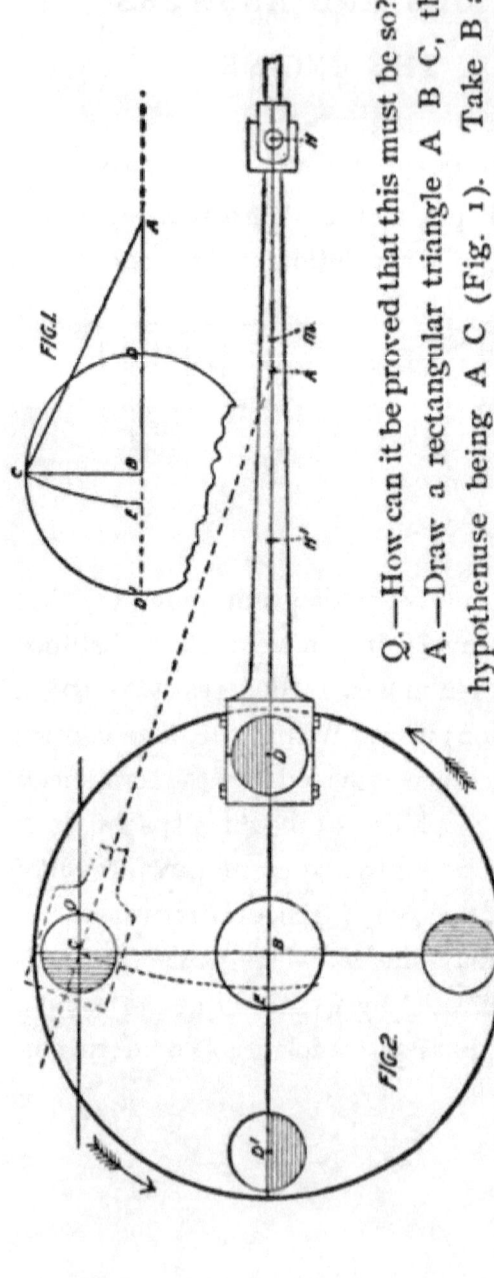

Q.—How can it be proved that this must be so?

A.—Draw a rectangular triangle A B C, the hypothenuse being A C (Fig. 1). Take B as the center of disc shaft, A as the center of crosshead pin, D D₁ as the dead centers, and C as the highest elevation of crank pin center. Then it is evident that the line A C is longer than A B, as a hypothenuse is longer than either of the two sides forming the right angle. Now, A C is the length of connecting rod from pin to pin, in the position of highest elevation (see Fig. 2), and by laying A C off on A B we find A C = A B + B E.

During one stroke the crank pin moves from D to D1 in a half-circle, and the crosshead moves an equal distance in a straight line, H H1. Therefore, the crosshead moves the distance H A during the first half-stroke, and the distance A H1 during the second half-stroke. M indicates the point dividing the length of stroke in halves.

Another proof of the difference is furnished by the length, which the central point of the connecting rod strap travels around the circumference of the crank pin. Comparing the two positions (in Fig. 2) at dead center and highest elevation, it is seen that o indicates one-half of the not shaded semicircle, around which the center of connecting rod strap has traveled during the half-stroke. But the dotted line showing the center of the connecting rod (in the moment of highest elevation) passes below the point o, which proves that more than one-half of the semicircle has passed the central point of connecting rod strap.

Q.—Why is it that the crosshead moves less than half its stroke during the second quarter of the circle traveled by the crank pin? Does not the crank pin travel as far as in the first quarter?

A.—Of course, it does; but in the first quarter it moved *away from* the level, besides the crosshead's motion along the level, and in the second quarter it moves *toward* the level.

MISCELLANEOUS

Q.—Is a heavy disc or a fly wheel of service in this connection?

A.—Yes, it serves to make the speed in a revolution more even, so that the shaft revolves steadily, while the unevenness of motion is put on the piston.

Q.—What other purpose does a fly wheel serve?

A.—It serves to overcome the dead centers, where the piston can neither push nor pull.

Q.—Does a crosshead stand still at dead center points?

A.—No. The crosshead center could stand still only if the crank pin moved around it as the center. The following cut shows how the reality differs from such a case.

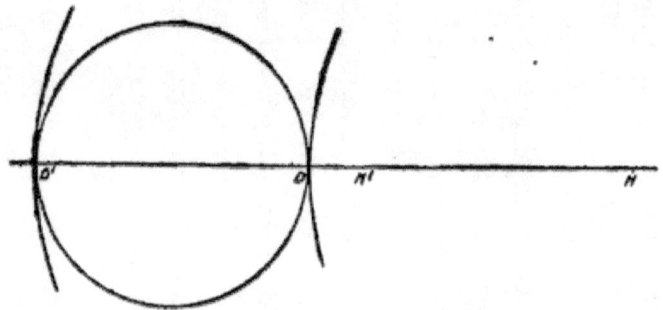

H, H1 are the positions of the crosshead pin center when the crank pin center is at D, D1. The circle shows the real movement of crank pin; the two curves indicate the circles in which the crank pin would have to travel as long as the crosshead stood still.

The movement near the dead centers is comparatively slow, but as the crank pin does not stand still at the dead center, but is moving either toward it or away from it, the crosshead, moving with it, does not stand still either. The dead center is an imaginary point, having no dimensions; thus it cannot be said that the crank pin center *remains* at the dead center point any *time*, or that it takes the crosshead any time to change its direction of stroke.

Q.—Can you illustrate this fact?

A.—Yes. The pendulum of a clock does not stand still at either end of its arc of oscillation. No time intervenes between the end of one year or month or hour and the beginning of the next.

Q.—Do the connecting rod brasses wear the crank pin evenly all around?

A.—No, in running "over" only one-half of the circumference of the crank pin is pushed and pulled, while in running under it is the other half. (In the full page cut the half affected in running "under" is shaded; the half affected in running "over," the direction indicated by the arrows, is not shaded.)

Q.—Why are horizontal engines (stationary) generally run over and not under?

A.—So the thrust will be downward upon the foundation rather than up against the caps of the boxes and the upper guides.

Q.—How much farther does the crank pin travel than the crosshead each revolution?

A.—One-half farther. The crosshead moves twice the diameter of the disc (back and forth), while the crank pin travels around the circumference ($= 3.1416$ times the diameter); 3 is more than 2 by half.

Q.—Does a crank pin have a tendency to flatten on one side, or on both sides, traveling in one direction?

A.—Simply on one side. The push and pull of the rod is on one-half of the pin only, as the pin turns with the crank, wheel or disc.

HEAT

Q.—Is heat a substance?

A.—No. Scientists say now it is the energy of molecular action.

Q.—What are molecules?

A.—The smallest possible parts into which any substance can be divided without losing its chemical identity.

Q.—What is meant by "absolute zero"?

A.—The absolute cessation of molecular action.

Q.—When heat is applied from outside to a substance, what are the effects?

A.—The substance increases in temperature, changes its volume, and, at certain degrees of temperature, changes its form. (See page 180.)

Q.—What is meant by "latent heat"?

A.—In expanding, gases take heat from their surroundings. This amount of heat does not increase the temperature of the expanding gas, and is therefore not measurable by the thermometer. Any heat expended in this or a similar way, and not "sensible," or "noticeable to the feeling," is called latent heat.

Q.—How is sensible heat measured?

A.—By means of a thermometer.

Q.—How is a **thermometer** constructed and graduated?

A.—A glass tube with bulb at closed end is partly filled with mercury, and heated until the mercury overflows. Then the open end is closed by fusing, and when cooled, the bulb is placed in melting ice and the point to which the mercury falls is marked the freezing point, 32 deg. Then place it in boiling water which is exposed to the open air and when the mercury rises to its full height, mark it 212 deg., or boiling point. The distance between the two points is 180 deg.

Q.—What causes the mercury to rise and fall?

A.—Expansion and contraction.

MEASUREMENTS AND CALCULATIONS

Q.—Give a general rule for determining the sizes of piston rods for steam engines?

A.—They should be 1-6 the diameter of the piston-head.

Q.—Does this rule answer for all-sized cylinders?

A.—No; only sizes ranging from 4 inches up to 28-inch cylinders. For sizes above 28 inches the piston rods are smaller in proportion.

Q.—Suppose there were 2 pounds of steam in the cylinder, how much pressure would there be between the piston head face, valve face and cylinder head? Explain by rule?

A.—Rule: First find the area of piston and multiply by pressure in cylinder.

Q.—What is the meaning of the term "clearance" in an engine cylinder?

A.—The unoccupied space between the valve face, cylinder head and piston head at each end of the stroke.

Q.—Which end of the cylinder has the most power?

A.—The end without the piston rod.

Q.—Explain why so?

A.—Because the steam has more square inches to act upon in the end without the piston rod.

Q.—How would you know the safe pressure to carry in a boiler ½ inch steel, 42 inches diameter, and 50,000 lbs. tensile strength?

A.—First multiply thickness of shell by full tensile strength and divide by half the diameter (radius), and divide by 6, which gives the safe pressure allowed by U. S. if welded. If boiler shell is double riveted multiply by .70 (= 70 per cent), in single riveted multiply by .56 (= .56 per cent).

Q.—What is understood by a unit?

A.—The basis of measurements, such as the day for measurements of time; the dollar for money; the atmosphere (14.7 lbs. per sq. inch) for pressure; the caloric (heat required for raising temperature of one lb. of water one degree) for heating; the horse-power (33,000 lbs. raised one foot high) for energy; the volt for electromotive force, etc.

Q.—What is a "thermal unit"?

A.—The amount of heat found necessary to raise or lower a pound of water 1 degree (Fahr.) of temperature.

Q.—What is meant by positive and negative heat?

A.—The former means the work of actual heating, the latter means the work done in cooling.

Q.—How is the weight of the atmosphere found?

A.—By the barometer.

Q.—How does it show?

A.—Air, being a substance, has weight. The atmosphere surrounding the earth presses at sea level with an average weight of 14.7 lbs. per sq. inch. This atmospheric pressure balances a column of mercury, in the vacuum arm of a siphon, of about 30 inches height. As the air rises (when heated, as in summer over a sandy plain) or sinks (when cold or heavy with moisture), we have lower or higher pressure and the barometer indicates this by the lower or higher position of the top of the mercury column in the vacuum tube. For better indication the scale is generally attached to the open arm, which is made very narrow so as to show greater differences.

Q.—How much does the whole atmosphere weigh?

A.—It is estimated at five trillions of tons, the weight of a solid leaden ball of 60 miles diameter.

Q.—How large should the stack be in proportion to the area of the tubes or flues combined of a stationary boiler?

A.—The stack should be about 25 per cent, or $\frac{1}{4}$ larger in area to do good work.

Q.—How many square feet of heating surface is generally allowed to 1 square foot of grate surface?

A.—From 22.5 to 40 square feet.

Q.—Suppose the area of a valve is known, how is the diameter found?

A.—Divide the area by .7854 and extract square root—answer equals diameter.

Q.—How is the radius (half diameter) found when area is known?

A.—Divide area by 3.1416 and extract square root—answer equals radius.

Q.—How is the linear dimension of a square found from the area?

A.—It is its square root. (See page 239.)

Q.—How much will 1 cubic inch of cold water expand when changing to steam?

A.—About 1728 times, or into 1 cubic foot.

Q.—How large should the diameter of a pump cylinder (plunger) be to deliver 324 gallons of water per minute, traveling 100 piston speed?

A.—Divide 324 by constant 4, equals 81; from this extract the square root—answer equals 9 inches, diameter of plunger.

Q.—What size should the steam cylinder be as compared with the pump cylinder?

A.—One-third larger in diameter. In the case mentioned, it should be 12 inches.

Q.—How much of the steam generated in a boiler is allowed for consumption in the engine?

A.—One-half only. With the usual average of 70 lbs. steam and the feed water at the temperature of 100° F., each 15 square feet of heating surface of the boiler will evaporate 30 lbs. of water per hour. The engine should, therefore, consume only 15 lbs. of water per hour for every 15 square feet of boiler-heating surface.

Q.—What if the boiler is not capable of generating double the amount of steam consumed by the engine?

A.—The boiler will be overworked, which means shortness of life, many repairs, a great waste of labor and fuel, and much annoyance.

MECHANICAL REFRIGERATION AND ICE MAKING

THE SCIENTIFIC PRINCIPLE

It is a well known fact that metals expand or contract as they are heated or cooled. Many other substances have the same quality, and it is a scientific truth, applying to all substances capable of expansion, that *a change of volume implies either an absorption of heat from, or a loss of heat to, the surroundings.*

Different substances have their extremes of volume at different temperatures. It is by no means so, that any substance will keep decreasing in volume indefinitely as its temperature decreases, or that its volume will keep increasing with increasing heat. Water, for instance, occupies the smallest space at 39.2° F. (the temperature found at the bottom of deep lakes), and has its extreme expansion at 212.8° F. (when it boils, or, in other words, when it changes from the liquid condition to the gaseous). Below 39.2° F. water *expands with decreasing temperature* (this is why ice floats and bursts pipes); and it cannot be heated beyond 212.8° F., except

REFRIGERATION AND ICE MAKING 181

in a closed vessel, which serves to *compress the steam into a smaller volume.*

From the above it will be understood that

Gases when compressed yield heat to, and when expanding absorb heat from, their surroundings.

THE APPLICATION OF THE PRINCIPLE

A gas which will rapidly increase its volume when surrounded by a very low temperature, taking the heat it needs for expansion from the surroundings, will, therefore, create around it a very cold region.

Scientists have frozen water in a bottle placed in a fire. The bottle was wrapped in woollen rags soaked in ether or chloroform, which evaporate so rapidly that they draw heat enough from the water in the bottle to freeze it.

Common air can be reduced to a liquid by the alternate application of enormous pressure and of cooling. Liquid air boils at 312° below 0 F., a temperature almost inconceivable. Undoubtedly, liquid air will become a mighty agent in the hand of man before long. (See page 223.)

Such a gas is, also, *anhydrous ammonia*, which boils under ordinary atmospheric pressure at 28.5° below zero F. By compressing it in strong steel tanks it is kept in a liquid condition, and is sold that way. From this tank (drum) it passes as a gas (vapor), feeding into a pump, which compresses it again. In this condition it enters the evaporating coils, in which it is allowed to expand rapidly. In this rapid expansion a consumption of heat is necessary, and the required heat is taken from the brine in which the coils are immersed. Then the expanded vapor is exhausted into a condensing tank, the evaporating coils receive a new charge of condensed vapor from the pump, and the operation is repeated. In this way the brine is kept at the desired low temperature. The brine cannot solidify (freeze) on account of the salt it holds in solution.

Q.—What is mechanical refrigeration?

A.—It is produced by the evaporation of a volatile liquid which boils at a low temperature, and which by means of evaporating coils, a condenser and a gas compressor, is brought under the control of the operator.

AMMONIA

Q.—What does the name "*anhydrous*" ammonia mean?

A.—"Anhydrous" means "free from water" or "dry."

Q.—What is ammonia, and where is it found?

A.—It is a gas composed of 1 part of nitrogen and 3 parts hydrogen. It can be obtained from the air, from sal-ammoniac, nitrogenous constituents of plants and animals by process of distillation. As a matter of fact, there are very few substances free from it. At present almost all the sal-ammoniac and ammonia liquors are prepared from ammoniacal liquid, a by-product obtained in the manufacture of coal gas and coke.

Q.—What are the properties of ammonia?

A.—Pure ammonia liquid is colorless, having a peculiar alkaline odor and caustic taste. It turns red litmus paper blue. Its boiling point depends on its purity, and is about 28½ deg. F. below zero at atmospheric pressure. The purer the liquid the lower its boiling point. Compared with water, its weight or specific gravity at 32 deg. F. is about ⅝ of water, or 0.625. One cubic foot of liquid ammonia weighs 39.73 lbs., 1 gallon weighs 5.3 lbs. One pound of the liquid at 32 deg. will occupy 21.017 cubic feet of space when evaporated at atmospheric pressure. The specific heat of ammonia gas (heat required to raise one unit of it one degree of temperature, as compared with the heat required for the same weight of water, =1.) is 0.50836. Its latent heat of evaporation, as determined by the highest authorities, is not far from 560 thermal units at 32 degrees.

Q.—What is a "refrigerant"?

A.—Anything that cools, such as **ammonia** known as anhydrous or dry ammonia.

Q.—What is the ammonia condenser?

A.—It is that part of the apparatus in which the gas is cooled and changed to a liquid.

Q.—How is the water changed into ice?

A.—By a system of evaporating coils in which the liquid ammonia is expanded into gas, thereby cooling the space around by absorption of the heat.

ICE MACHINE. BEAM PATTERN—CORLISS ENGINE.

Q.—At what degree does pure anhydrous ammonia boil?

A.—At from 28½ to 40 deg. below zero. (See table of boiling points, page 210.)

Q.—What advantage does this give?

A.—Ammonia can be kept at its boiling point without any artificial heat, which is not possible with water.

Q.—Is ammonia the most serviceable of all refrigerants?

A.—Yes, it has many advantages over other refrigerants.

Q.—Which standard is applied to the amount of ammonia consumed in producing cold, weight or volume?

A.—Weight.

Q.—Is ammonia inflammable and explosive?

A.—It is not inflammable, and is, therefore, not explosive in the sense in which gunpowder is explosive; but at any temperature above 28.5 below zero F. it is expansive like dry steam, and is, therefore, dangerous.

Q.—Has ammonia any corrosive effect on steel or iron?

A.—No; but on brass it eats.

Q.—Has it any effect when mixed with water on the machinery or piping?

A.—No.

ABSORPTION METHOD AND COMPRESSION METHOD

Q.—Can you describe the **absorption method** or system?

A.—Yes, but it is very little used now.

The gas, instead of being compressed by mechanical means, is obtained from a 26 per cent solution of ammonia in water, heated in a boiler or still, until the ammoniacal gas is driven off. This gas then goes through the cycle of operations as described, until, having done its work of refrigeration, it is conveyed into the **absorber**.

Here the gas is brought in contact with the water, called the mother liquid, from which it was

originally extracted in the still, this water in the meantime having gone through an elaborate process of cooling. The cool mother liquid rapidly absorbs the gas and forms again a strong solution of ammonia. This solution is returned to the still by means of a pump and is ready again to go through the same cycle (round) of operations.

Q.—Name the parts of an absorption apparatus?

A.—Generator, ammonia pump, absorber, condensing tank, weak liquor tank, equalizer, freezing tank, cooling tank and receiver for ammonia.

Q.—What is the absorption system based upon?

A.—The chemical law which allows ammonia to boil into gas at 28.5 deg. below zero, while water is not affected until 212 deg. is reached. By this the ammonia and water are capable of being separated and made to perform continuous duty.

Q.—Is the **compression system** based on the same difference?

A.—No, because anhydrous ammonia is used in this system. ("Anhydrous" means *without water*.)

Q.—Why is it called a compression system?

A.—Because it consists of alternate compression and expansion of the refrigerant.

Q.—What are the different operations in this system?

A.—There are three, namely, 1st, compression of the gas; 2d, condensation of the gas and a withdrawal of the heat caused by compression; 3d, expansion of the gas and absorption by it of the heat from the surrounding objects.

Q.—Explain process of **compression?**

A.—The refrigerating agent (anhydrous ammonia) is furnished in heavy iron drums and allowed to enter, through connecting coils, the induction pipe in the compression pump, from whence it is drawn into the cylinders, where it is compressed to a pressure varying from 125 to 175 lbs. per sq. inch. This variation of pressure is regulated by the temperature of the condensing water. This compression produces, by a largely increased friction of the gas molecules (small particles), intense heat.

Q.—Does the pump get hot?

A.—Yes, cold water is kept flowing around it, to cool it.

Q.—Explain process of **condensation?**

A.—The compressed gas is then allowed to enter the system of pipes known as the condenser, over which cold water is kept constantly flowing (see cut, page 203). The cold water absorbs the heat generated in the process of compression. The gas is thus cooled in its flow through the great lengths of pipe, until it finally cools to below 28.5° F., when it collects in the receiver as a liquid.

Q.—Explain process of **expansion?**

A.—The liquefied ammonia, through a siphon, now slowly enters the expansion or evaporating coils, which are brought in contact with, or in close proximity to the objects to be cooled. As it enters these coils the high pressure before mentioned is reduced, and the ammonia immediately

ABSORPTION AND COMPRESSION METHODS 189

re-expands into the gaseous condition, absorbing the heat necessary for this process from the pipes and through them from the surroundings. Wherever two bodies of different temperature are brought in contact, the hotter will impart its heat to the colder until the temperatures are equalized.

Q.—After having thus accomplished its cooling work, where does the gas go?

A.—It is returned to the compressor, there to again begin afresh the cycle of operations, namely, compression, condensation and expansion.

Q.—Is there any **loss of ammonia** during each operation?

A.—Yes, very small.

Q.—How often can ammonia be used in the manner just described?

A.—Times without number.

Q.—What is absolutely necessary to render these three operations continuous?

A.—Each separate part of the machine (apparatus) must be suitably connected. (See testing and charging.)

Q.—State the main points as to all appliances and machinery about a refrigerating plant?

A.—Good order and cleanliness should be practiced, also pump and valve tightly packed.

Q.—What gives the greatest trouble about an artificial ice plant?

A.—Leakage.

BRINE SYSTEM AND DIRECT EXPANSION SYSTEM

Q.—In what different ways is refrigeration done?

A.—For lesser degrees of cooling, as for breweries, living-rooms, etc., the **brine system** is sufficient, in which the brine after being cooled by the ammonia is pumped through the pipes. For very low temperature, as needed in cold storage, **direct expansion** is used, allowing the gas to expand in the pipes, which are placed in the cooling rooms.

Q.—Which of the two systems is more expensive?

A.—The direct expansion system, both because of the large amount of specially made pipe required, and because the whole plant must be in operation day and night, to supply liquid ammonia for expansion.

Q.—What advantage has the brine system over the direct expansion in ordinary conditions?

A.—Ordinary piping may be used, and the large body of brine suffices to maintain the temperature desired in the rooms for a considerable length of time by merely operating the brine circulating pump, it very frequently being only necessary to operate the compressor in the daytime to maintain the temperature during the entire twenty-four hours.

Q.—Describe the process of **circulation in the brine system**?

DIRECT EXPANSION SYSTEM

A.—It is done by a special pump known as the brine circulating pump, which forces it through the pipes arranged in the rooms to be cooled, from which it returns to the re-cooling tank and is used continually over and over again.

Q.—Is the brine circulation independent of the gas?

A.—Yes.

Q.—Where and when do they come in contact?

A.—In the brine tank only.

Q.—Explain how this is done?

A.—The cold ammonia gas extracts the heat from the brine as it flows through the tank in the circulation pipes.

Q.—Do the two circulating systems come any nearer than that just mentioned?

A.—No.

Q.—What is the brine tank in an ICE-MAKING PLANT?

A.—It consists of one or more salt water tanks, in which the evaporating coils of pipe are submerged, and the liquid ammonia is allowed to expand within, where it assumes its original gaseous condition and in so doing absorbs the heat from the brine, lowering the temperature to any degree required.

Q.—How is the brine tank arranged for making ice?

A.—It is a covered tank with many openings to

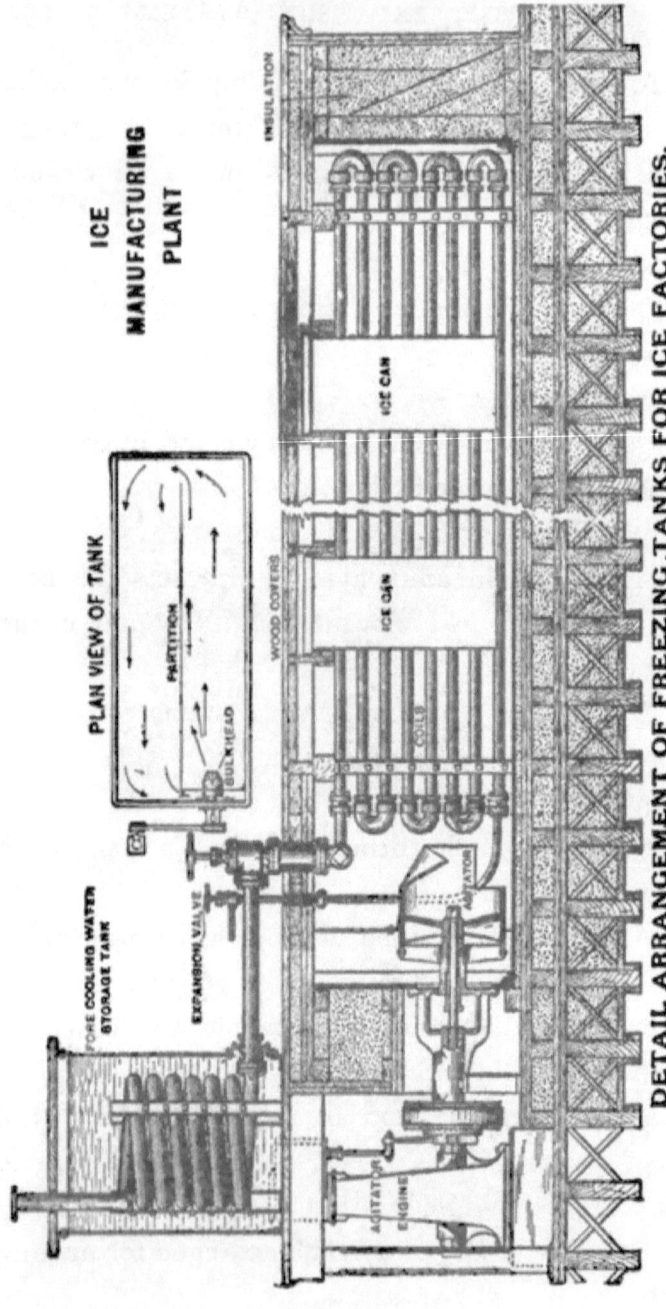

admit galvanized sheet iron tanks to hold distilled water for freezing into blocks of clear ice. (See opposite page; for model ice plant, see page 10.)

Q.—How long will it take the water in the galvanized tanks to freeze a cake of the usual size, 11x22x45 inches?

A.—It is according to the temperature of the brine.

Q.—If the brine is cooled to 14 deg. above zero, how long would it take?

A.—About 60 hours.

Q —Why does it take so long?

A.—Ice is a bad conductor; the ice forming first on the six surfaces communicates the cold very slowly to the water within.

Q.—Is it a good plan to freeze the water quickly?

A.—No; if frozen too quickly it will not be transparent, but cloudy.

Q.—What is meant by the agitator and its use?

A.—It is a centrifugal (rotary) pump used for drawing the brine from the bottom of one end of the tank and discharging it in the other end at the top, thereby securing uniform freezing by continual circulation of the bath.

Q.—How is the ice cake taken from the mold?

A.—By running hot water over the can and dumping it. (See cuts of Eclipse thawing apparatus, next page.)

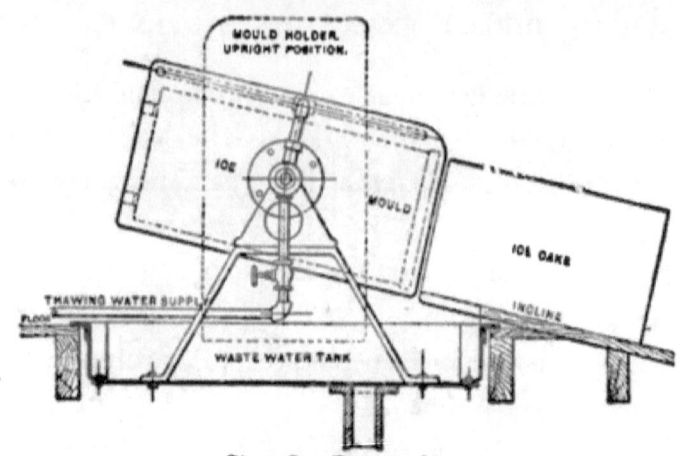

Plain Can Dump, of iron

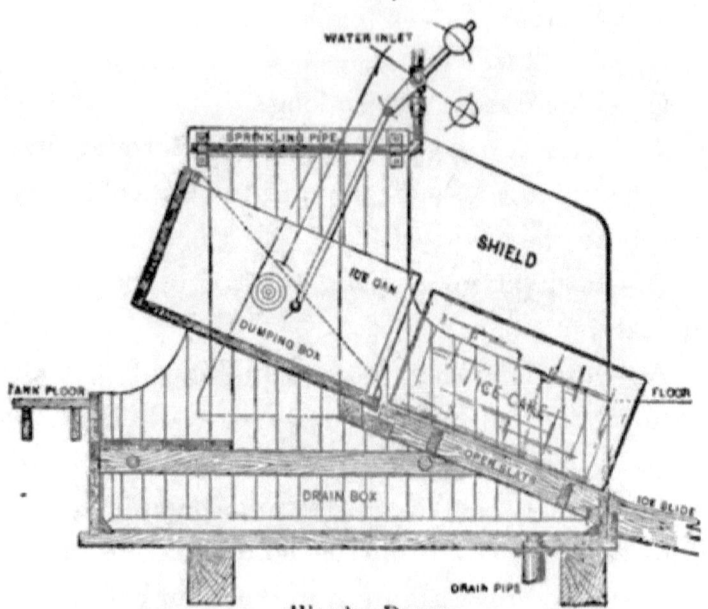

Wooden Dump
ECLIPSE AUTOMATIC THAWING APPARATUS AND CAN DUMP.

TABLE OF BRINE SOLUTIONS
(Chloride of Sodium, Common Salt)

Percentage of Salt by Weight..........	0	1	5	10	15	20	25
Degrees by Salometer at 60° F	0	4	20	40	60	80	100
Specific Gravity at 60° F.............	1	1.007	1.037	1.057	1.115	1.156	1.191
Weight of One Gallon..............	8.35	8.4	8.65	8.95	9.3	9.6	9.94
Pounds of Salt in One Gallon.........	0	0.084	0.432	0.895	1.395	1.92	2.485
Pounds of Water in One Gallon........	8.35	8.316	8.218	8.055	7.905	7.68	7.455
Pounds of Water in One Cubic Foot...	62.4	62.172	61.465	60.253	59.124	57.408	55.695
Freezing Point in Degrees F.........	32	31.8	25.4	18.6	12.2	6.86	1.00

WATER EXAMINATIONS

FOR HARD WATER, ETC.

Particles suspended in the water may be detected by filling a tall glass cylinder and placing same on a clean piece of white paper and looking down through the water. All waters are known as hard or soft, and in many cases hard water may be made soft and the air and gases be expelled by boiling, in which case it is called temporary or removable hardness. If unaffected by boiling it is called permanent hardness, and nothing short of distillation or boiling into steam **and condensing the vapor will remove the cause.**

Q.—What is it that makes water hard, and how much of it is present?

A.—Eight grains of mineral matter (carbonate of lime, etc.) or more in a gallon of water make it hard.

Q.—How is the quality of hardness particularly noticeable?

A.—When **soap** is used, the harder the water, the less effect has the soap, because the mineral matter neutralizes so much of it.

SIMPLE RULES FOR ASCERTAINING THE QUALITY OF SO-CALLED MINERAL WATERS

Water which will turn blue litmus paper red before boiling, but not after boiling, is carbonated (contains carbonic acid). The blue color can be restored by warming.

If it has a sickening odor, giving a black sediment, acetate of lead, it is sulphurous (containing sulphureted hydrogen).

If it gives blue settlings with yellow or red prussiate of potash by adding a few drops of hydrochloric acid, it is chalytate (carbonate of iron).

If it restores blue color to litmus paper after boiling, it is alkaline.

If it has none of the foregoing properties in a marked degree and leaves a large residue after boiling, it is **saline water** (containing salts).

THE APPARATUS USED IN THE BRINE AND IN THE DIRECT EXPANSION SYSTEMS

THE PUMP VALVE

The induction or suction valve is shown closed, the piston being on its upward stroke. Surrounding the upper portion of the valve stem is seen a coiled spring which raises the valve, holding it firmly upon its seat, as shown above and in sectional view of compressor, page 199.

As the piston commences its downward stroke the pressure of the gas in chamber D opens the valve and the cylinder commences to fill.

Below A in Fig. 1 is seen a small passageway connecting the gas inlet space on the right with a small chamber on its left formed by the ring B on the valve stem and the bore of the valve cage. This passage opens a little above the bottom of the chamber, and when the valve is fully opened the ring B covers the passage, and the gas in the lower portion of the chamber, unable to escape, forms an elastic cushion, which prevents any strain on the valve stem and holds the valve in perfect equilibrium.

The downward stroke being complete, the incoming gas no longer presses open the valve and by the combined action of the spring and the imprisoned cushioning gas it is instantly seated.

The discharge valve is side by side with the induction valve, and works in the opposite sense.

The requirements of a good pump are: To instantly admit the gas to the cylinder, filling it full at each downward stroke of the piston; to expel (discharge) the entire contents of the pump through the outlet valve K, Fig. 2, which opens as soon as the cylinder pressure overcomes the combined force of the valve-spring and of the pressure in the condenser beyond the valve

Valves, of ample area, durable in construction and reliable in action, must be supplied. A piston is required that is perfectly tight, yet

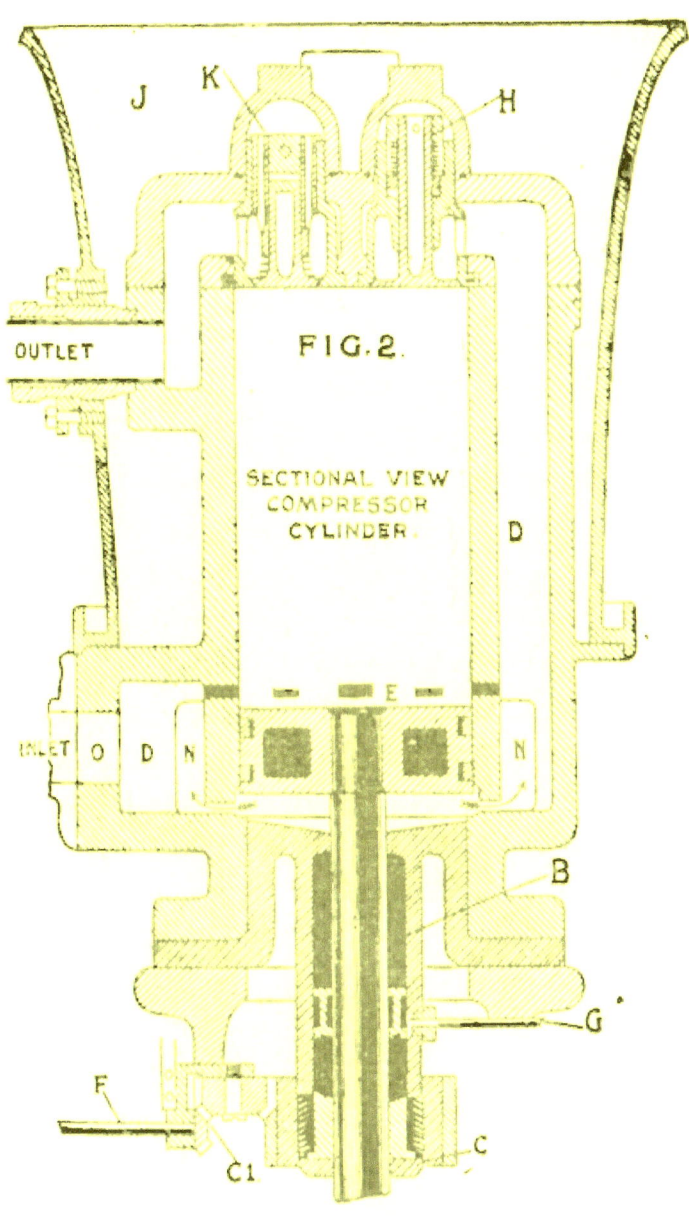

working freely, and a stuffing box for the piston rod in which the packing can be readily adjusted while in operation.

The stroke of the piston is accurately gauged so as to reach within a hair's breadth of the upper cylinder head in order to force all the gas out.

Just above the lowest position of the upper face of the piston head there is a ring of 8 openings in the cylinder wall, connecting with the induction chamber D. Through these holes, whose total area equals that of the induction valve, the gas enters at the lowest position of the piston head, thus securing a complete filling of the cylinder, after the induction valve has closed. For the induction valve, as shown by the indicator, admits only about three-fourths of the desired amount of pressure, because, with the spring tension, this is enough to balance the pressure in the induction chamber.

As the piston rises again, it closes the ring of 8 openings, until it passes beyond them, when the gas enters once more through them to fill the vacuum under the piston. In the downward stroke, when the piston closes these openings, the remaining gas under it is pressed through small openings (shown white in the cut) at the bottom of the cylinder into the closed chamber N, whence it issues again at the beginning of the upward stroke, working like a cushion.

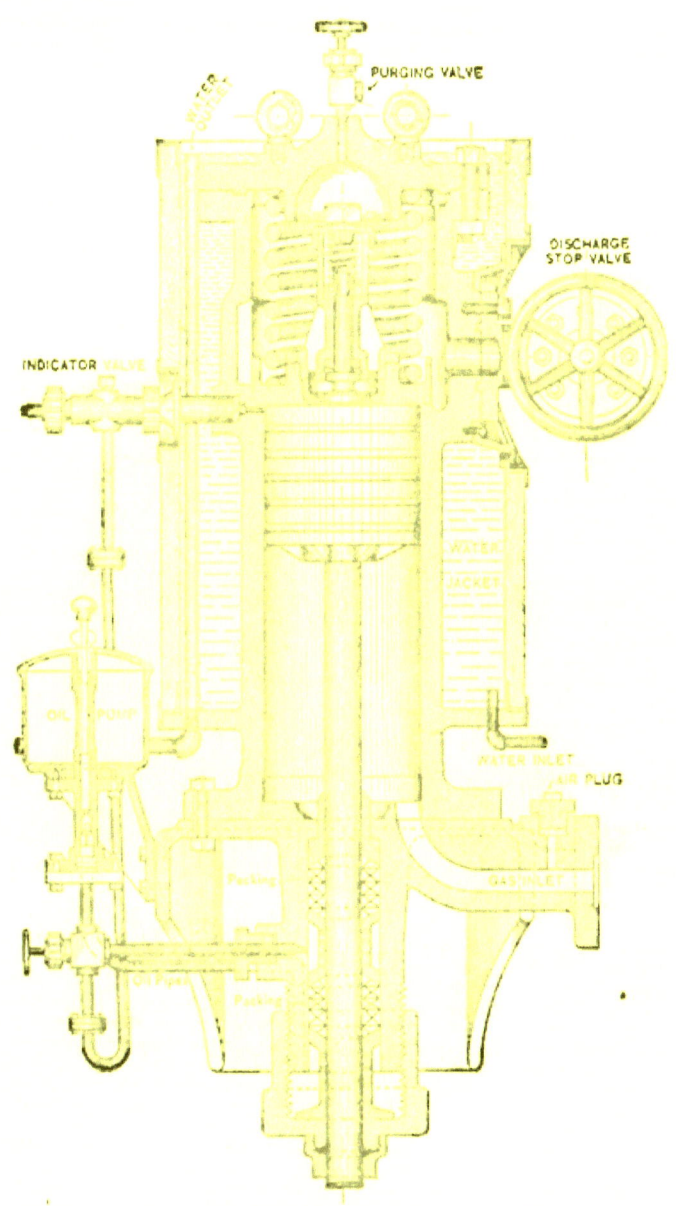

AMMONIA COMPRESSION PUMP

THE STUFFING OR PACKING BOX

The leakage of ammonia, even if so slight as to cause but little expense, is always an annoyance. Confined as it is in the pipe system, in endless coils without the possibility of escape, the only portions of the plant needing careful attention, to guard against leaks, are the stuffing boxes B, Fig. 2, of the compression pump piston rods A.

The stuffing box is of unusual depth, but with whatever care it is designed, engineers are aware that frequent attention is required in all machines to keep the packing set up enough to prevent leakage, and still not so much as to induce heating and the consequent cutting of the rods.

The stuffing box is under perfect control of the engineer at all times. Its geared gland C_1, Fig. 2, connects with a short rod F. Turning the handle F will tighten or loosen the packing. The engineer can regulate the pressure upon the packing while the pump is in motion.

To guard against the leakage of ammonia, in addition to the very long stuffing box already mentioned, a lubricating chamber with oil pipe G is attached for lubricating the piston rod within the packing.

THE AMMONIA CONDENSER

The ammonia—leaving the compression pumps hot, compressed, but still gaseous—reaches the condenser, which consists wholly of piping and should be conveniently located on the roof of the building. The condenser should be divided into two parts, namely, the preliminary condenser and the liquefier, as shown in the illustration on this page.

The gas when discharged from the compressor

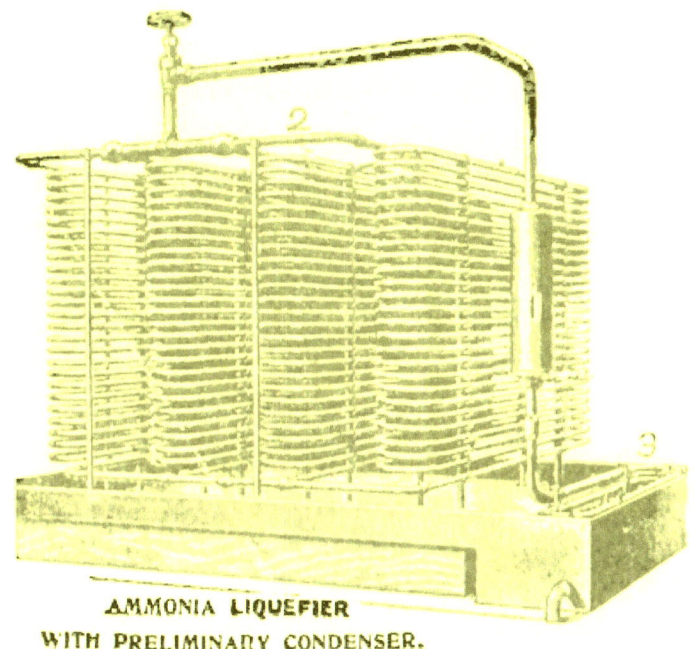

AMMONIA LIQUEFIER
WITH PRELIMINARY CONDENSER.

passes into a trap where oil and other foreign matters are deposited; from the trap it passes into the preliminary condenser (3), which is located a little lower than the bottom of the liquefier. After being cooled to a considerable extent in the preliminary condenser the gas passes through another oil trap (1), which thoroughly eliminates even the slightest trace of oil still remaining in the ammonia. When it is remembered that the ammonia supply ought to do its work for a long period of time in the performance of the never-ending cycle of operations, and that all foreign substances act injuriously on both gas and machinery, it is apparent that it is of the most vital importance to keep the gas absolutely free from all impurities. This the additional oil trap successfully accomplishes.

Thus, completely purified, the gas passes out to the liquefier (2), where it is cooled to a liquid.

The condensing pipes are cooled by water. In the **open air system** the water drips on the top coils and from them down on the lower ones, until it reaches the shallow tank in which the preliminary condensing coils are immersed. Thus the water is hottest when it meets the hottest gases. A waste pipe carries the overflow of hot water to the steam condenser.

In another cooling system, the condensing pipes are entirely **submerged in a water tank**, the water flowing in at the bottom and running out near its surface. The work of condensing can, therefore, not be divided up as in the open air system.

The divided form of condenser possesses marked advantages and is a great improvement over the old method of arrangement. The warmest water meets the hottest gas, and, as it has already performed duty on the liquefier, it is used on the preliminary condenser, without expense.

All the coils should be made from extra heavy special drawn pipe, bent cold, and finished coils, which should be tested under many times the pressure they will ever be subjected to in actual use.

The liquid ammonia flows into the receiver, where it is ready to perform the work of cooling either by expanding into coils in tanks or by expanding into coils in rooms to be cooled.

THE EXPANSION COILS

The ammonia, which left the compression pumps and entered the condenser as a gas through a large pipe, now leaves the condenser in a pipe from one-half to one inch in diameter, and enters the third division of the system, there again to expand into its original gaseous condition. And it is while expanding to this gaseous condition that the ammonia absorbs the heat from the surrounding objects ere it returns to the compressor to be again compressed.

SOLDER AND SOLDERING FLUID

Bar solder: 1 lb. block tin, ½ lb. lead.
Glazing solder: 1 lb. block tin, 1 lb. lead.
Plumbing solder: 1 lb. block tin, 2 lbs. lead.
For a good soldering fluid, drop small zinc strips into 1 oz. of muriatic acid until the bubbles cease to rise, then add ¼ teaspoonful of sal ammoniac.

COLD STORAGE TEMPERATURES

ARTICLES.	°Fahr.	ARTICLES.	°Fahr.
FRUITS.		**CANNED GOODS.**	
Apples	32-36	Sardines	35-40
Bananas	54	Fruits (Nuts in shell)	35-40
Berries, fresh	36	Meats	35-40
Cranberries	33-36		
Cantaloupes	40	**BUTTER, EGGS, ETC.**	
Dates, Figs, etc.	50-55		
Fruits, dried	35-40	Butter	18-20
Grapes	34-36	Butterine	18-20
Lemons	33-36	Cheese, Chestnuts	34
Oranges	34-36	Eggs	31
Peaches	34-36		
Pears, Watermelons	34-36		
		LIQUIDS.	
MEATS.		Beer, Ale, Porter, etc.	33
Brined	38	Cider	30
Beef, fresh	33	Ginger Ale	36
Beef, dried	36-40	Wines	40-45
Calves	32-33		
Hams, Ribs, Shoulders (not brined)	20	**FLOUR AND MEAL.**	
Hogs	29-32	Buckwheat, — Wheat Flour	36-40
Lard	38	Corn Meal, Oats	36-40
Livers	20-30		
Sheep, Lambs	32		
Ox-tails	30	**MISCELLANEOUS.**	
Sausage Casings	20		
Tenderloins, Butts, etc.	33	Furs, Woolens, Cigars, etc.	35
VEGETABLES.		Honey, Maple Syrup, Sugar	40-45
Asparagus, Carrots	34-35	Hops	40
Cabbage, Celery	34-35	Oils	35
Potatoes	36-40	Poultry, dressed, iced	28-30
Dried Beans, Corn, Peas	32-40	" dry picked	26-28
Onions, Parsnips	34-35	" scalded	20
FISH.		Game — Poultry — to freeze	15-18
Fresh Fish	20	Game, after frozen	25-28
Dried Fish	36		
Oysters in shell	30-35		
Oysters in tubs	25		

AMMONIA VALVES AND FITTINGS

1. Expansion valve.
2. Ninety-degree angle valve.
3. Cross valve.
4. Return bend.
5. Ammonia tank valve.
6. Forty-five-degree angle valve.
7. Coupling.
8. Joint cross.
9. Liquid valve.
10. Ninety-degree angle purger valve.
11. Union loop.

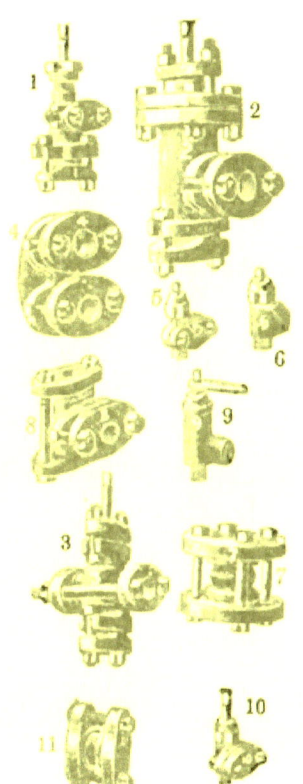

BYE PASS VALVE ON THE DOUBLE AMMONIA GAS PUMP

Through the bye pass the ammonia can be readily exhausted from any part of the system and may be stored in any other part temporarily until the repairs or examinations are made.

By the peculiar arrangement of pipes and valves the action of the compressor and pump can

208 BYE PASS VALVE

be reversed and the gas pumped from the condenser, storing it in the brine tank.

In each case, after the examination of any part, the air can be exhausted therefrom and charge of ammonia reintroduced without the admixture of air.

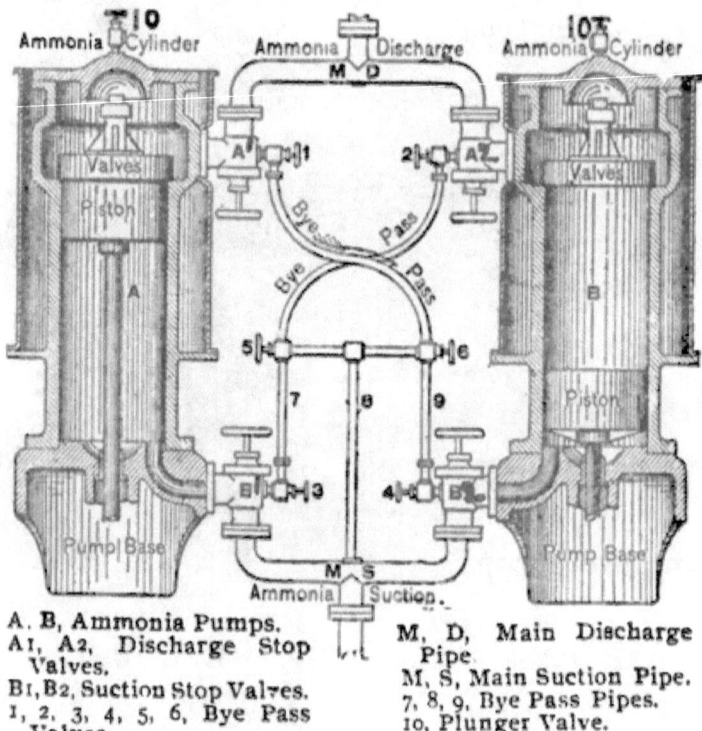

A, B, Ammonia Pumps.
A1, A2, Discharge Stop Valves.
B1, B2, Suction Stop Valves.
1, 2, 3, 4, 5, 6, Bye Pass Valves.
M, D, Main Discharge Pipe.
M, S, Main Suction Pipe.
7, 8, 9, Bye Pass Pipes.
10, Plunger Valve.

DESCRIPTION: A, B, compressor pumps; A1, A2, main discharge stop-valves; B1, B2, main suction stop-valves; 1, 2, 3, 4, 5, 6, bye pass valves; M, D, main discharge pipe; M, S, main suction pipe; 7, 8, 9, bye pass pipes.

HOW TO OPERATE: To exhaust gas from pump

3, all bye pass valves should be closed to begin with; close main stop-valve B1, B2 and A2; open bye pass valves 2 and 3; then by running pump slowly the contents of pump B can be exhausted; then close valve 4 and remove bonnet. After closing bonnet, air can be removed in same way, previously shutting main stop-valve A1 and expelling the air through purging valve on pumphead; close all bye pass valves when done and open main stop-valve.

To EXHAUST PUMP A—Proceed in same manner, using the opposite set of valves.

To EQUALIZE PRESSURE between condenser and brine tanks—Open stop-valve A1 or A2 and bye pass valves 1 and 2, also 5 and 6, thus forming passage direct from main discharge to main suction pipe.

To EXHAUST CONDENSER and store gas in brine tank—All valves closed to begin with. Open stop-valve A1 on pump A, bye pass valves 1 and 4, opening communication to pump suction B; expel gas by opening bye pass valves 2 and 5, thus discharging into main suction pipe. Run pumps slowly by using opposite set of valves (either pump may be used), the mode of operation being simply that one pump is used to exhaust the gas through the bye pass from the discharge while the other forces it through the other half of bye pass into the suction pipe.

BOILING POINT OF AMMONIA

Pressure.		Boiling Point °Fahr.	Latent Heat.	Pressure.		Boiling Point	Latent Heat.
Absolute.	Gauge.			Absolute.	Gauge.		
10.69	−4.01	−40	579.7	58.00	43.30	28.9	537.6
11.00	−3.70	−39	579.1	59.41	44.71	30.0	536.9
12.31	−2.39	−35	576.7	60.00	45.30	30.6	536.5
13.00	−1.70	−32.7	575.3	61.50	46.80	32.0	535.7
14.13	−0.57	−30	573.7	62.00	47.30	32.3	535.5
14.70	∓0.00	−28.5	572.3	63.00	48.30	33.0	535.0
15.00	+0.30	−27.8	571.7	64.00	49.30	33.7	534.6
16.17	1.47	−25	570.7	65.93	51.23	35.0	533.8
16.71	2.01	−22	568.9	67.00	52.30	35.8	533.3
17.00	2.30	−21.8	568.7	69.00	54.30	37.2	532.4
18.45	3.75	−20	567.7	71.00	56.30	38.6	531.5
19.00	4.30	−18.9	567.0	73.00	58.30	40.0	530.6
20.99	6.29	−15	564.6	74.07	59.37	41.0	530.0
21.27	6.57	−13	563.4	75.00	60.30	41.5	529.7
22.10	7.40	−12	562.8	76.00	61.30	42.2	529.2
22.93	8.23	−11	562.2	78.00	63.30	43.4	528.5
23.77	9.07	−10	561.6	80.66	65.96	45.0	527.5
24.56	9.86	−9	561.0	88.96	74.26	50.0	524.3
25.32	10.62	−8	560.4	92.00	77.30	51.4	523.4
26.08	11.38	−7	559.8	95.00	80.30	53.2	522.3
26.84	12.14	−6	559.2	97.93	83.23	55.0	521.1
27.57	12.87	−5	558.5	100.00	85.30	56.1	520.4
28.09	13.39	−4	557.9	104.84	90.14	59.0	518.6
28.64	13.94	−3	557.3	107.60	92.90	60.0	517.9
29.17	14.47	−2	556.7	110.00	95.30	61.1	517.2
29.70	15.06	−1	556.1	115.00	100.30	63.5	515.7
30.37	15.67	∓0 zero)	555.5	118.03	103.33	65.0	515.3
31.00	16.30	+1.4	554.6	119.70	105.00	66.0	514.8

A FEW TESTS FOR AMMONIA

Ammonia liquid for use in refrigerating machines should be absolutely pure. It should be tested. The various tests to which it should be subjected are: For water, for specific gravity, for inflammable gases, and for boiling point.

TEST FOR WATER

As shown in the engraving, screw into the ammonia flask a piece of bent ¼-inch pipe, which will allow a small bottle to be placed so as

TESTING FOR WATER BY EVAPORATION.

to receive the discharge from it. This test bottle should be of thin glass with wide neck, so that quarter-inch pipe can pass readily into it, and of about 200 cubic centimeters capacity—equals 1.69 gills or a 6¾ ounce bottle. Put the wrench on the valve and tap it gently with a hammer. Fill the bottle about one-third full and throw sample out in order to purge (clean) valve, pipe and bottle. Quickly wipe off the moisture that has accumulated on the pipe, replace the bottle and open valve gently, filling it about half-full. This last operation should not occupy more than one minute. Remove the bottle at once and insert in its neck a stopper with a vent hole for the escape of the gas. A rubber stopper with a glass tube is the best, but a rough wooden stopper loosely put in will answer the purpose. Procure a piece of solid iron that should not weigh less than 8 lbs., pour a little water on this and place the bottle on the wet place. The ammonia will at once begin to boil and in warm weather will soon evaporate. If any residuum, pour it out gently, counting the drops carefully. Sixteen drops are about equal to one cubic centimeter, and if the sample taken amounted to 100 cubic centimeters, sixteen drops of residuum shows one per cent impurities (adulteration), and 20 drops 1¼ per cent.

Care is necessary in the drawing of the sample, as a very little moisture in the bottle, or in the pipe, or a brief exposure to the atmosphere will at once affect its purity.

TEST FOR SPECIFIC GRAVITY

The specific gravities of aqua ammonia by the Beaume scale are given in the following table.

AMMONIA TESTS

By drawing off some of the liquid in the tall test tube generally provided by ice-machine builders, the Beaume hydrometer may be inserted and the specific gravity read upon the scale. If water is present, the liquid will show a density proportionate to the percentage of water present.

TABLE OF SPECIFIC GRAVITIES AND PERCENTAGE OF AMMONIA (CARIUS)

Degrees Beaume.	Specific Gravity.	Percentage.	Degrees Beaume.	Specific Gravity.	Percentage.
10	1.000	0.	21	.9271	19.4
11	.9929	1.8	22	.921	21.4
12	.9859	3.3	23	.915	23.1
13	.9790	5.	24	.909	25.3
14	.9722	6.7	25	.9032	27.7
15	.9655	8.4	26*	.8974	30.1
16	.9589	10.	27	.8917	32.5
17	.9524	11.9	28	.886	35.2
18	.9459	13.7	29	.8805	
19	.9396	15.5	30	.875	
20	.9333	17.4			

*Called by the trade 29½ per cent.

Specific Gravity of pure anhydrous ammonia is .623

TEST FOR INFLAMMABLE GASES

Take a pail of water, submerge the bent pipe therein, open the valve on flask slightly and allow a small quantity of gas to flow into the water. If inflammable gases are present they will rise in bubbles to the surface of the water and may be removed by igniting the bubbles by means of a lighted match or candle. As water has a strong affinity for ammonia it will be readily absorbed, while air or other gases will show only in the form of bubbles.

TEST FOR BOILING POINT OF ANHYDROUS AMMONIA

By inserting the special low temperature standardized chemical thermometer into liquid drawn into the 6¾ oz. test glass jar, readings can be obtained through the side of the jar without removing the instrument. Hold the thermometer in such a position that only the bulb is immersed.

This test will give you the boiling point of ammonia at atmospheric pressure and it is well to know that the state of the barometer affects the temperature of the boiling point. With the barometer at 29.92 inches the boiling point should not be above 28.6 deg. below zero and may be much lower, depending upon purity of sample. If the ammonia is impure the boiling point is raised in proportion.

TESTING THE REFRIGERATING MACHINERY

PRESSURE TEST.—It is important before introducing the charge of gas into the machine system to carefully test every part of the apparatus and make it thoroughly tight under at least 300 lbs. air **pressure**, which pressure may be obtained by working the ammonia compressor (pump) and allowing free air to flow into suction side of pump by opening special valves generally provided for the purpose, the entire system being thus filled with compressed air at the desired pressure.

While this pressure is being maintained a search is instituted for leaks, every pipe, joint and square inch of surface being scrupulously noted. One method is to cover all surfaces with a thick lather of soap, leaks showing themselves by formation of soap bubbles. In the case of condenser and brine tank coils, the tanks are allowed to fill with water, the bubbles of air escaping through the water locating the leak.

It is important that the apparatus be thoroughly tight, and, as a few joints are to be made when new plants are put in, it is necessary to go over the entire surface of the system to be sure.

While the machine is engaged in pumping air into the system advantage should always be taken of this opportunity to purge (clean) the system of all dirt and moisture. To do this properly, valves are provided so the apparatus may be blown out by sections, removing valve covers (bonnets), loosening joints for this purpose, so that it is positively known that each pipe, valve and space is strictly clean and purged of all dirt and traces of moisture.

A final test may then be had by pumping air pressure of 300 lbs. into the entire system and allowing the apparatus to stand for some hours, estimating the leakage, if any, by noting the degrees of pressure as shown by the pressure gauge connected to system. The air pressure will shrink somewhat at first, by reason of losing heat gained during compression by the pumps. As soon as the air parts with its heat and returns to its normal temperature, the gauge will come to a standstill and remain at a fixed point (depending upon the barometer and upon the temperature of the room), if the system is tight. Never charge a system until it is well cleansed, purged and absolutely tight.

VACUUM TEST

After having tested the system with a pressure of 300 lbs. of compressed air, the air must be exhausted from the entire system, by working the pumps and discharging the air through valves provided therefor (located generally on the pump domes). When the escape of air ceases and the compound or vacuum gauges show a full vacuum, it is well to close all outlets and allow the machinery and system to stand for some time, to test the capacity of the apparatus to withstand external pressure without leakage. In some cases it has been discovered that parts while tight from internal pressure, owing to loose particles lodging over leaks and acting as plugs to prevent escape, give way and disclose the leakage when subjected to an external pressure.

INTRODUCING THE CHARGE OF AMMONIA

Place the ammonia flask (tank) on small platform scales, in order to weigh the contents and know positively when flask is exhausted. Connect the flask to the charging valve, the gauge still showing a vacuum, close the expansion valve in main liquid pipe connecting receiver to brine tanks; then open valve on ammonia flask and allow the liquid to be exhausted into the system.

The machinery may be run all this time at a slow speed, with both discharge and suction hand stop valves wide open.

As one flask is exhausted, place another on the scales and continue until the liquid receiver is shown to be partly full by the glass gauge thereon. Then shut the charging valve and open and regulate the main expansion valve. The machine is then sufficiently charged to do work, as shown by the pressure gauges and gradual cooling of the brine and frosting of expansion pipe leading to brine tank coils.

While the system is being charged water is allowed to flow on the condenser, and time diligently employed in searching further for leaks, which can readily be detected by sense of smell, each joint being again gone over.

Q.—Why are the joints and whole system again gone over after having withstood the two tests, 300 lbs. air pressure and lowest vacuum?

A.—Because ammonia in itself is a great dissolvent and eventually it will purge and scour the entire system clean to the metal surfaces.

Q.—Where does the loose foreign matter go?

A.—It is caught in the separators and intercepters provided for this purpose.

Q.—Is ammonia a lubricant?

A.—Yes, slightly.

Q.—Has it any effect on iron or steel?

A.—None whatever.

Q.—Has it any effect on brass, copper, etc.?

A.—Yes, it eats and corrodes them.

AIR IN THE SYSTEM

Q.—What causes air to get in the system?

A.—Negligence in regulating the expansion valve, needlessly pumping a vacuum on the brine tank, leaky piston rods, also taking the apparatus apart and not expelling the air before the reintroduction of the anhydrous ammonia gas.

Q.—How is the pressure of air in system in considerable quantity readily noticed?

A.—By the intermittent action of the expansion valve and singing noise, rise of condensing pressure, loss of efficiency in the condenser, etc.

Q.—What means are provided for the escape of

DISCHARGING AMMONIA

the imprisoned air to restore the apparatus to its normal condition of pressure and efficiency?

A.—The purging (cleaning) valves on the condenser or the bye pass. (See description of bye pass on pumps, pages 208 and 209.)

TAKING THE AMMONIA OUT OF THE APPARATUS

Q.—How can ammonia be taken out of the system of an ice machine without losing any of it?

A.—If the plant is of ordinary construction and of compression (liquefying gas by gas pump) design, connect the liquid receiver by its bottom connection to the empty shipping tank and allow the gas to flow into the tank, being sure to have the tank on a scale to weigh the quantity you put in.

Do not allow more to be placed in the tank than was originally in it when shipped. In other words, the tank must not be filled with liquid to more than five-eighths of its cubic contents.

This is one of the most dangerous pieces of work that a refrigerating engineer is called on to do, and on the first trial the chances are about even that he will burst the compressor, blow the receiving tank up and possibly blow his own head off.

QUESTIONS AND ANSWERS IN REVIEW

Q.—How is the compressor pump cylinder kept cool?

A.—It is incased with a water jacket through

which cold water is constantly circulated (see J. Fig. 2, page 199).

Q.—What causes the heat in the pump?

A.—The compression of the gas.

Q.—What kind of oil should be used in the compressor, if used at all?

A.—Oil generally known as "the perfection ammonia pump oil," or the cold test "zero oil," which is especially manufactured, and which stands a very low degree of cold without volatilizing. Sometimes the best paraffine oil is used, and again a clear West Virginia crude oil. These oils when subjected to a low temperature should not freeze.

Never inject oil directly into the compressor, and use sparingly in the stuffing box.

Q.—What is an oil separator used for?

A.—It is to eliminate the small quantity of oil from the ammonia gas in its passage from the compressor to the condenser.

Q.—Is the ammonia gas, when exhausted, inflammable?

A.—Yes, sometimes, if the oil traps have not absorbed the oil which the gas carries off from the hot pump.

Q.—Is ammonia dangerous to handle?

A.—It is, because when condensed to a liquid it is under an enormous pressure, which may cause great destruction when suddenly released.

Q.—What advantage has the bye pass valve?

A.—By means of it the ammonia can be exhausted from any part of the machine that may need repairing.

Q.—What is an ammonia receiver, and where is it placed?

A.—It is a tank to store liquefied ammonia, and is placed between the condenser and expansion valve (see condenser).

Q.—What do the pipes in a cold storage room with a very low temperature contain?

A.—They contain ammonia gas.

Q.—What do the pipes used in a hotel for cooling living rooms contain?

A.—They contain brine.

Q.—What does a gas need for expansion and how does it get what it needs?

A.—It needs heat and takes it from the surroundings.

Q.—But what, if the surroundings have no heat?

A.—Heat means any degree of temperature. Taking heat from surroundings means lowering their temperature. Taking heat from cold surroundings means making them still colder.

Q.—In which three forms does matter exist?

A.—Solid, liquid and gaseous.

Q.—Does iron exist in these three forms?

A.—We can liquefy it by melting it in great

heat, and it is affirmed that iron exists in gaseous form in the sun.

Q.—How can this be known?

A.—"Spectral analysis" reveals the fact.

Q.—Why does ice float?

A.—Because ice is lighter than water at any temperature.

Q.—What makes it lighter?

A.—Expansion. A pound of ice has more volume than a pound of water.

Q.—What would happen, if ice were denser and therefore heavier than water at any temperature?

A.—In severe winters the deepest lakes would freeze solid down to the bottom.

Q.—Does ice keep on expanding, the colder it grows?

A.—No, there is a point at which it begins to contract again.

Q.—Why does it not keep the same volume?

A.—Because change of temperature is impossible without change of volume.

Q.—Is that a law of nature?

A.—It is a truth established by sufficient observation, that one never occurs without the other.

A STEAM AND WATER-PIPE CEMENT

that will set under water, is made of 2 lbs. ground Paris white, 5 lbs. ground lithage, ¼ lb. fine yellow ochre, ¼ oz. hemp cut up small. Mix well with linseed oil to the consistence of putty, and use at once.

LIQUID AIR, THE COMING FORCE

Water freezes at 32° above zero. Mercury in a thermometer freezes solid at 40-42° below zero. The alcohol in a spirit thermometer freezes at 200 below. Air becomes a liquid at 312° below zero.

Eight hundred cubic feet of free air are condensed into one cubic foot of liquid air. One pint of liquid air weighs one pound, like water.

By the aid of a 50-horse-power steam air pump ordinary air is compressed until it becomes red-hot. Then it is cooled in submerged pipes, and is further compressed until the pressure is registered at thousands of pounds to the square inch. More cooling is done, and more pressure applied, until, finally, the air liquefies. It oozes through the steel of the pipe in the shape of a milky vapor and trickles down into the receptacle below.

As there is a difference of 344° between the temperatures of ice and liquid air, it will be understood why liquid air boils furiously even when placed on a block of ice.

A hand thrust into this liquid, in appearance like water, would be destroyed in 10 seconds, but if drawn out again instantly, the moisture of the skin freezing to ice would be protection enough. The feeling at touching the liquid is like that of iron at white heat.

Like quicksilver, liquid air does not adhere. If poured over silk, it will leave no stain.

When boiling, the vapor of liquid air, being nothing but highly-compressed air, sinks to the ground.

If water is poured into liquid air it turns to ice instantly, and of such a low temperature that it will not melt near a red-hot stove for a long time.

A stick of arc light carbon, heated to 2,000 degrees above zero, thrust into liquid air, causes the oxygen in it to burn with a dazzling bright flame.

A teaspoonful of liquid air in a closed vessel, if lighted, explodes with tremendous force, jarring the ground like an earthquake.

The expansive power of liquid air is about 20 times greater than that of steam.

Ten years ago it cost about $2,000 to produce 1 gallon of liquid air. To-day, so Prof. Chas. E. Tripler, of New York, states, it can be manufactured at the cost of 3 or 4 cents a gallon, at the rate of 40 or 50 gallons a day.

Some of the uses to which this uncanny substance can be put are as follows:

A steam engine horse-power is now figured at $36.00 a year expense; by the use of liquid air it should not be more than about $7.00.

The resistance in electric wires is entirely overcome, if submerged in liquid air. The intense cold knits the molecules of metal so closely that it becomes a perfect conductor, without any leakage.

A pocket flask full of liquid air will furnish free air for submarine apparatus for hours.

It furnishes a clean, dry cold, at any desired temperature, for refrigerators, hospitals, engine rooms, etc.

If used in propelling steamships, there would be no heat in the furnace room, and little need of a furnace.

Guns using liquid air as an explosive would never get hot.

Liquid air sprayed on dangerous wounds arrests blood poisoning instantly, as by a miracle. Malignant cancers have been cured by one drop of the liquid. All pulmonary and throat diseases, hay fever, asthma, diphtheria, grip and all fevers yield to a spray of liquid air.

LIQUID HYDROGEN

In the spring of 1898, Prof. Dewar, of the British Royal Institution, succeeded in liquefying the most volatile of all gases, hydrogen. Liquid hydrogen is colorless, transparent, and of only one-fourteenth of the density of water. It is so cold that it freezes and solidifies air and oxygen instantly. In a closed tube brought in contact with it, the air freezes into a small lump, leaving the tube a vacuum.

THE MACHINE SHOP

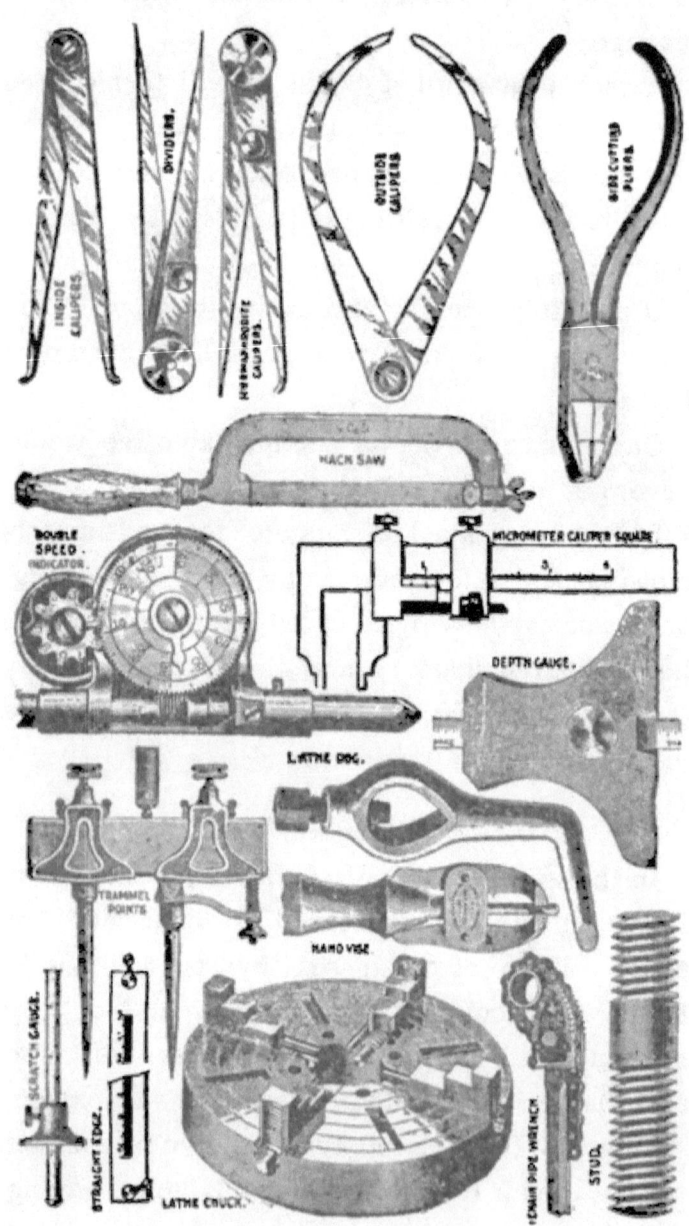

TOOLS 227

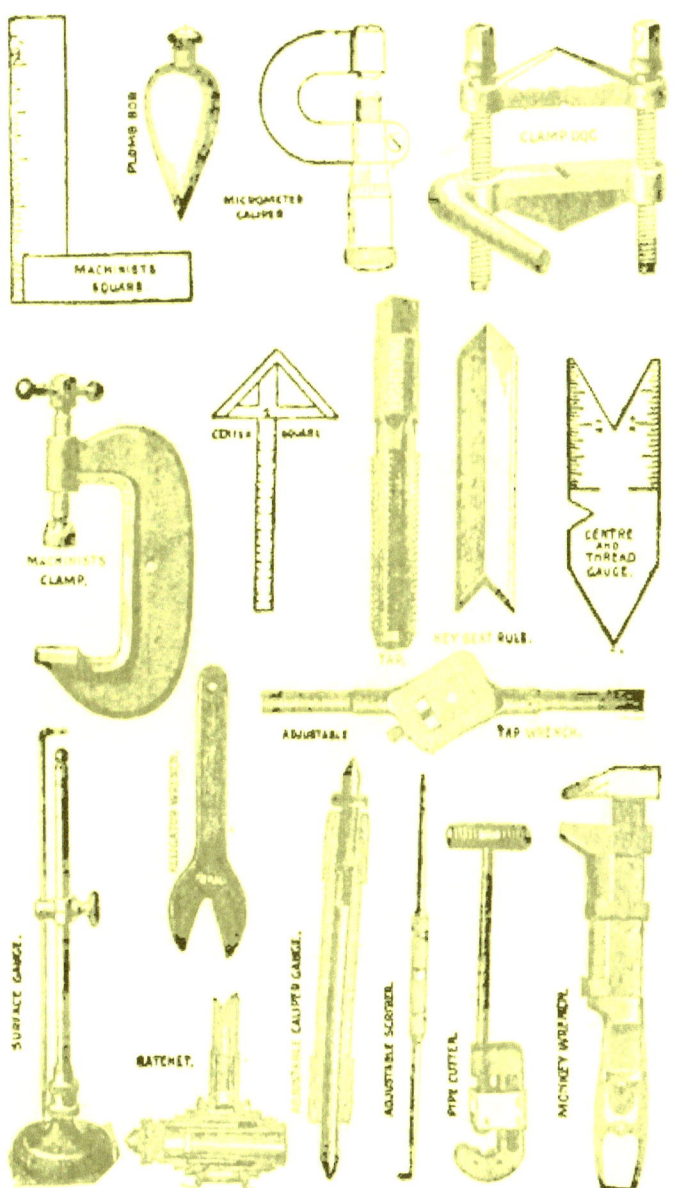

THE MACHINE SHOP

One of the things by which a mechanic is known, is the way he keeps his tools. It makes no difference whether he works in a small shop or in one of the great establishments, every mechanic should be **inflexible** in following these

TWO RULES:

1. Every tool should have its exact place, and should be in that place when not in actual use.
2. Every tool should be in good order and ready for use.

A mechanic with whom the constant observation of these rules has grown to be a habit, is worth three others to his employer, and saves himself a great amount of annoyance, loss and worry.

LATHE GEARING

To gear a lathe to cut any number of threads **when no gear** plate is attached **to lathe head**

block, simply find the run of gears belonging to the lathe to know if odd or even number of teeth are on gears.

Multiply the number of threads to be cut to the inch by any small number from 3 up to 6 that will bring the answer even with one of the gears on hand. Say 10 threads are to be cut—4 times 10 equals 40. Place this gear on lead screw of lathe. Multiply the same number (4) by the number of threads per inch on your lead screw, say 6. 6x4= 24. Place 24 tooth gear on spindle, and connect by suitable intermediate.

TURNING A BALL.

There are expensive machines for turning balls, but common lathes will produce perfect spheres.

Turn the piece first on centers, using the calipers to get it approximately near the shape; then cut off the centers.

Make a cup in a chuckblock of hardwood, to hold a small section of the ball, and for the center use a blunt wood center with a concave piece of copper. Put the work in the chuck so as to take the first cut around it in the direction of its former centers, or axis.

Cut lightly and a very narrow ribbon all around; then change the chuck so as to cut the second ribbon at right angles with the first, with the same depth of cutting. Then the third

LATHE TOOLS

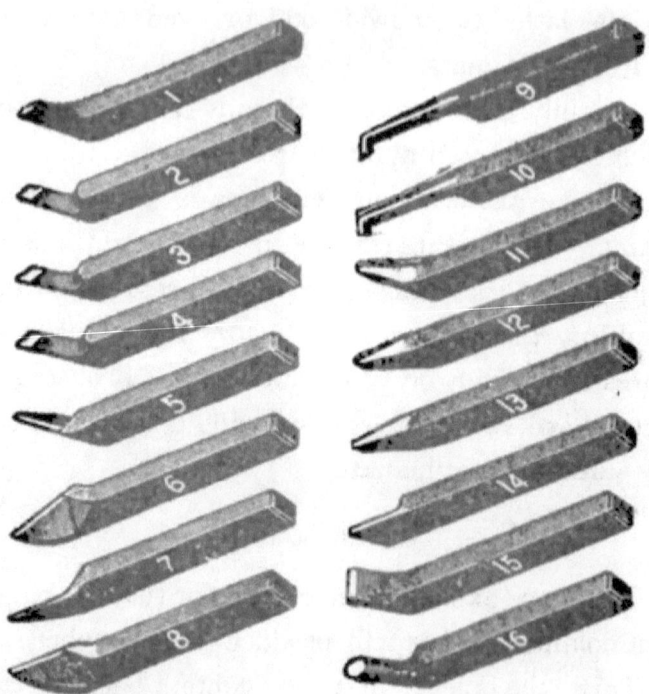

1. Half Diamond Point.
2. Diamond Point for steel and iron, left hand.
3. Diamond Point for steel and iron, right hand.
4. Heavy Diamond Point for Cast Iron.
5. Right Side Tool, bent.
6. Left Side Tool, bent.
7. Right Side Tool.
8. Left Side Tool.
9. Inside Thread Tool.
10. Inside Turning (Boring) Tool.
11. Bent Thread Tool.
12. Thread Tool, straight.
13. Roughing Tool.
14. Cutting-off Tool.
15. Water Finishing Tool.
16. Round Nose Tool.

NOTE: Set the cutting edge a little above the axis, or it will not cut properly, and may be drawn under and broken off.

ribbon half-way between the first two, and so on, until the whole surface is covered. The right angle need not be measured except with the eye.

Finish the ball with a hand tool, or scraper.

TWIST DRILL GRINDING

The cutting edges of a drill must have a proper and uniform **angle** with the longitudinal axis of the drill (Fig. 1); the two edges must be straight and exactly of the same **length**; and the lips must be sufficiently backed off (Fig. 6).

Q.—What is the proper angle to which a drill should be ground?

A.—59 degrees. (See Fig. 1.)

Q.—What is the result of an improper angle?

A.—A lesser angle gives a longer edge, likely to hook and to produce a crooked and irregular hole. A larger angle gives too short an edge to do the work easily.

Fig. 1

Q.—Where is the longitudinal axis of the drill?

A.—It is at the intersection of the two longitudinal planes indicated by the scribing (center line) along the middle of the two grooves.

Q.—Of what importance is this axis?

A.—The cutting edges must be at equal distances from it, and also at the right distance to get the *proper angle of point*.

Q.—What is this point?

A.—It is the part where the two edges of the lips are run together in the center. If the cutting edges are too far from the axis the angle point does not cut; if too near, it cuts too rank.

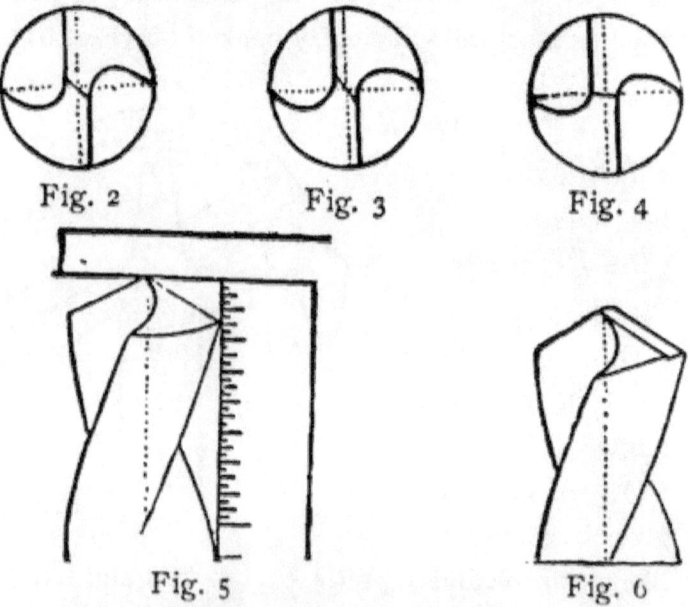

Fig. 2   Fig. 3   Fig. 4

Fig. 5   Fig. 6

Fig. 2 shows the proper proportions. In Fig. 3 the edge is too near the center line, and in Fig. 4 it is too far from it.

Q.—What is clearance?

A.—The amount which is champered off back from the cutting edge.

Fig. 5 shows how the clearance is determined as well as the height of the cutting lips, which should be equal, as stated before.

Q.—What, if there is not sufficient clearance?

A.—The drill will not cut, and under force will split or break. Fig. 6 shows the rear of lip removed.

Q.—How would you start a drill?

A.—By hand, in order to see first how it works. If it cuts well, the chips will show a clean cutting surface.

Q.—Does a good drill in the machine cut small chips?

A.—In cast metal, yes; but in wrought metal, it will cut a curled shaving sometimes very long.

Q.—Why are the two grooves shallower near the shank than near the point?

A.—The center is made thicker toward the shank for strength. As the drill wears short, the center must be thinned out by grinding, care being taken to remove an equal amount of stock on each side and so keep the point central.

POLYGONAL NUTS

A 4 sided nut is called square.
A 5 " " " " a Pentagon
A 6 " " " " " Hexagon.
A 7 " " " " " Heptagon.
An 8 " " " " an Octagon.

RULES AND STANDARD NUMBERS

DIAMETER, CIRCUMFERENCE AND AREA

In a circle of one inch diameter describe 16 radii at equal distances (Fig. 1). The spaces

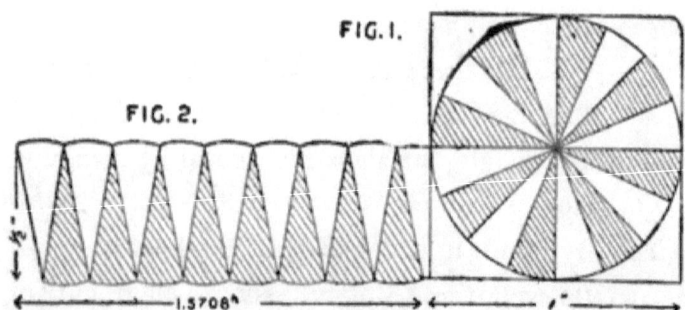

between them and the 16 parts of the circumference may be arranged in a double row (Fig. 2).

The circle area is thus divided up into 16 parts, 8 of which are placed in nearly a straight line on each side of the row.

By actual measurement the width of this row is ½ inch and the length 1.5708″ (allowing a trifle for the difference between the 8 little curves and a straight line).

Therefore circumference $= 2 \times 1.5708'' = 3.1416''$

Area $= 1.5708'' \times \tfrac{1}{2}'' = 0.7854$ sq. inch.

Difference of areas of circle and square $=$ sum of 4 corners (Fig. 1) $= 0.2146$ sq. inch.

RULES AND STANDARD NUMBERS 235

STANDARD MULTIPLIERS

1. For the area of a circle, multiply square of diameter by............. .7854
2. For the circumference of a circle, multiply diameter by................ 3.1416
3. For the diameter of a circle, multiply circumference by .... .31831
4. For the surface of a ball, multiply square of diameter by............ 3.1416
5. For the cubic inches in ball, multiply cube of diameter by .... .5236
6. For the cubic contents of a cylinder, multiply the area by the length.
7. For the pressure in lbs. per sq. inch in a column of water, multiply its height in feet by................... .434

AREA OF CIRCLES

Diam.	Area	Diam.	Area	Diam.	Area	Diam.	Area
⅞	0.6013	13	132.73	36	1017.8	71	3959.2
1	0.7854	14	153.93	37	1075.2	72	4071.5
½	1.767	14½	165.13	41	1320.2	76	4536.4
2	3.141	16½	205.97	45	1590.4	80	5026.5
¼	3.976	18	254.46	46	1661.9	81	5153.0
½	4.908	½	268.80	47	1734.9	82	5281.0
¾	5.939	19	283.52	48	1809.5	83	5410.6
3	7.068	½	298.64	49	1885.7	84	5541.7
¼	8.295	20	314.16	50	1963.5	85	5674.5
½	9.621	½	330.06	51	2042.8	86	5808.8
¾	11.044	21	346.36	52	2123.7	87	5944.6
4	12.566	½	363.05	53	2206.1	88	6082.1
½	15.904	22	380.13	54	2290.2	89	6221.1
5	19.635	½	397.60	55	2375.8	90	6361.7

Q.—What difference is there between 3 square feet and 3 feet square?

A.—The first means 3 squares, each one foot square; the second is 9 squares, each one foot square, arranged in 3 rows of 3 squares each.

Q.—Is there any difference between one square foot and one foot square?

A.—No.

WEIGHTS AND MEASURES

Q.—How many square inches in a square foot?

A.—12 times 12, or 144.

Q.—How many cubic inches in a cubic foot?

A.—12 times 12 times 12, or 1728.

Q.—How many cubic inches in a gallon, in a cubic foot, in a bushel?

A.—231 in a gallon; 1,728 in a cubic foot; 2,150 in a bushel.

Q.—How many gallons in a cubic foot of water?

A.—7½ gallons.

Q.—How many cubic inches in one pound of water at 60° F.?

A.—27.71 cubic inches.

Q.—How do you figure the gallons contained in a barrel?

A.—Add together the two diameters (in inches) of the barrel at head and bung, and divide the sum by 2, which gives the mean diameter. Multiply the square of this diameter by .7854, which gives the area of the mean diameter circle in sq. inches. Multiply this area by the length of the barrel in inches, to get the cubic contents in cubic inches, and divide the product by 231 to get the gallons.

Example: A barrel 40 inches long, 19 inches diameter at the head, 25 inches diameter at the bung.

```
19 + 25 = 44        44 ÷ 2 = 22
22 × 22 = 484       484 × .7854 = 380
380 × 40 = 15200    15200 ÷ 231 = 65.9 gals
```

Q.—How much does a cubic inch of water weigh?

A.—It weighs .0361 of a pound, or .577 of an ounce.

Q.—What is the weight of a gallon, a cubic foot of water?

A.—A gallon weighs 8⅓ lbs., a cubic foot 62½ lbs.

Q.—What is the weight of a column of water, one inch sq. and 2.309 feet high, the temperature at 60° F.?

A.—One pound.

Q.—What is the weight of a column of water, one inch sq. and one foot high?

A.—It weighs .434 lbs.

Q.—How much does a cubic inch of mercury weigh?

A.—It weighs .49 of a pound.

Q.—How much does a column of mercury one inch sq. and 30 inches high, weigh?

A.—14.7 lbs.

Q.—What is meant by a *miner's inch?*

A.—It is approximately equal to a supply of 12 gallons per minute.

Q.—In what relation does the friction of water in pipes stand to the velocity of flow?

A.—It increases with the square of velocity. If the velocity increases 4 times, the friction increases 16 times.

Q.—In what relation does the capacity of pipes stand to their diameter?

A.—It increases with its square. Doubling the diameter increases the capacity four times.

Q.—How much water is consumed in obtaining one nominal horse power in heating buildings, etc.?

A.—One cubic foot.

Q.—How much for engine purposes?

A.—One-half cubic foot.

Q.—How much heating surface is allowed for one nominal H. P. in boilers?

A.—15 sq. feet for horizontal, and 12 sq. feet for vertical.

Q.—How do you find the H. P. required to elevate water to a given height?

A.—Multiply the total weight of water in lbs. with the height in feet and divide the product by 33,000. Then allow 25 per cent for water friction and 25 per cent for steam loss, in all 50 per cent, or one-half, which is the same as dividing by 16,500 instead of by 33,000.

WEIGHTS AND MEASURES   239

Q.—How do you find the total amount of pressure exerted by a pump, and how the resistance?

A.—The area of the steam piston, multiplied by the steam pressure, gives the pressure. The area of the water piston, multiplied by the water pressure per sq. in., gives the resistance.

Q.—If pressure and resistance are the same, does the pump work?

A.—No, there must be a margin of from 30 to 50 per cent steam pressure according to the required speed.

PULLEY SPEED CALCULATION

Driven pulley revolutions are found by multiplying the diameter of the driver by its number of revolutions and dividing by the diameter of the driven.

Diameter of driving pulley is found by multiplying the diameter of the driven by the number of revolutions it shall make and dividing the answer by revolutions of driver per minute.

Diameter of driven pulley that should make a certain number of revolutions is found by multiplying the diameter of the driver by its number of revolutions and dividing by the revolutions the driven should make.

SQUARE ROOT

Q.—How is the square root of a number found?

A.—1st—Separate the number into periods of

two figures each, beginning at the right hand or digit space.

2d—Find the greatest number whose square is contained in the period on the left; this will be the first figure in the root.

3d—Subtract the square of this figure from the period on the left, and to the remainder annex the next period of two figures to form a dividend.

4th—Divide this dividend, leaving out the last single figure on the right, by double the part of root already found, annex the answer to that part and also to the divisor, then multiply the divisor thus completed by the figure of the root last obtained and subtract the product from the dividend.

5th—If there are any more periods to be brought down continue the operation in the same manner as before.

Note—If a cipher occurs in the root, annex a cipher to the trial divisor and another to the dividend, and proceed as before.

EXAMPLES:

```
 18,66,24 | 432 Sq. Root.          .0,00,36 | .006 Sq. Root.
 16                                 0
 ──                                 ──
 83 | 266                           00 | 00
      249                                00
      ───                                ──
 862 | 1724                        006 | 0036
       1724                              0036
       ────                              ────
```

LEVERAGE

Q.—Name the three points in a lever.

A.—Force, weight and fulcrum.

Q.—If the fulcrum is between the force and weight, what kind of a lever would it be?

A.—A lever of the first kind.

Q.—If the weight is between the force and the fulcrum, what kind would it be?

A.—A lever of the second kind.

Q.—When the force is between the weight and the fulcrum, what kind would it be?

A.—It is a lever of the third kind.

Q.—State how the proportions of a lever of the first kind are found?

A.—By dividing the length of that end of the lever between fulcrum and weight into the length of the opposite end. Example: If length of lever between fulcrum and weight is 6 inches and the other 18 inches the lever is said to be 3 to 1, and a weight equal to 3 times the force applied at force may be lifted at weight by pulling down at force.

Q.—How do you figure the lever of the second kind?

A.—By dividing the length of the end of lever between the fulcrum and weight into the total length of the lever. Example: If length of lever between fulcrum and weight is 6 inches and the other 24 inches the lever is said to be 4 to 1, and a weight may be lifted at weight equal to 4 times the force applied.

Q.—What are the lever proportions of third kind?

A.—They are found by dividing total length of lever into the length of the end between fulcrum and force. Example: If total length of lever is 30 inches and the length between weight and force 24 inches, the lever is said to be an 8-10 to 1, and a weight equal to 8-10 of the force applied at force may be lifted at weight.

GENERAL USEFUL KNOWLEDGE

AIR PURIFIER FOR ENGINE ROOM AND MACHINE SHOP

The contrivance consists of a tubular casing adapted for insertion in a circular opening in the roof of a building or the deck of a vessel. Inside of the casing a sleeve is so supported as to leave an air-passage between the casing and the sleeve. Mounted in the sleeve is a tube provided internally with a spider or frame, and at its upper end with a rotatable ingress tube. This ingress tube likewise has a spider or frame on which a rod is centrally pivoted. The upper end of the casing is inclosed by a hood formed with a conical end, through which the ingress tube passes. With the conical end of the hood an ingress tube is connected which communicates with the interior of the hood. These ingress and egress tubes are curved in opposite directions, and are mounted to swing in such a manner that the ingress tube shall constantly present its opening to the wind.

The ingress tube continually forces a column of air downward through the building, and the egress tube permits all warm or vitiated air to escape. Any vacuum formed by ventilation, it is said, will be immediately filled by the air pressed into the cold tube entering a room at the bottom. The ventilator at the rear or leeward of the hood constitutes an air-passage, creating a vacuum below and drawing up the warm air.

HOW TO READ A GAS METER

The right hand dial of the three used for actual measurement, records the number of feet by hundreds, up to 1,000, the center dial the number of thousands up to 10,000, and the left hand one the number in tens of thousands up to 100,000. Thus, if the hands have passed the 5, 6 and 7 figures on these dials the amount consumed is 76,500, etc.

THERMOMETERS

Comparative Scales.			Rules for Conversion.
Reaumur, 80°.	Centigrade, 100°.	Fahrenheit, 212°.	
76	95	203	Abbreviations: F. = Fahrenheit, C = Centigrade, R. = Reaumur
72	90	194	
68	85	185	
63.1	78.9	174	
60	75	167	
56	70	158	
52	65	149	
48	60	140	
44	55	131	
42.2	52.8	127	
40	50	122	
36	45	113	
33.8	42.2	108	**To Convert**
32	40	104	
29.3	36.7	98	F. to C., subtract 32 and multiply remainder by $\frac{5}{9}$.
28	35	95	
25.8	32.2	90	
24	30	86	F. to R., subtract 32 and multiply remainder by $\frac{4}{9}$.
21.3	26.7	80	
20	25	77	
16	20	68	
12.4	15.3	60	C. to F., multiply by $\frac{9}{5}$ and add 32.
10.2	12.8	55	
8	10	50	
6.8	7.2	45	R. to F., multiply by $\frac{9}{4}$ and add 32.
4	5	41	
1.3	1.7	35	C. to R., multiply by $\frac{4}{5}$.
0	0	32	
— 0.9	— 1.1	30	R. to C., multiply by $\frac{5}{4}$.
— 4	— 5	23	
— 5.3	— 6.7	20	
— 8	—10	14	
— 9.8	—12.2	10	
—12	—15	5	
—14.2	—17.8	0	
—16	—20	— 4	
—20	—25	—13	
—24	—30	—22	
—28	—35	—31	
—32	—40	—40	

STOPPING WITH A HEAVY FIRE

When it becomes necessary to stop an engine with a heavy fire in the furnace, place a layer of fresh coal on the fire, shut the damper and start the injector or pump for the purpose of keeping up the circulation in the boiler.

TO PREVENT ACCIDENT BY THE SHAFTING

While the shafts are in motion, it is strictly prohibited: *a.* To approach them with waste or rags, in order to clean them. *b.* To climb upon a ladder or other convenience in order to clean a shaft.

These parts of the machinery must be cleaned by means of a long-handled brush only, and while standing upon the floor.

The workmen charged with these or other functions about the shafting must wear jackets with tight sleeves, and closely buttoned up; they must wear neither aprons nor neckties with loose ends.

Driving pulleys, couplings and bearings are to be cleaned only when at rest.

This labor should, in general, be performed only after the close of the day's work. If performed during the time of an accidental idleness of the machinery, or during the time of rest, or in the morning before the commencement of work, the engineer in charge is to be informed.

GRAPHITE IN STEAM-FITTING

The value of graphite in making joints cannot be overestimated. Indestructible under all changes of temperature, a perfect lubricant and an anti-incrustator, any joint can be made up perfectly tight with it and can be taken apart years after as easily as put together. Rubber or metal gaskets, when previously smeared with it, will last almost any length of time, and will leave the surface perfectly clean and bright. Few engineers put to sea without a good supply of this valuable mineral, while it seems to be almost overlooked on shore.

HOW TO OVERCOME VIBRATION

How to put the smith shop in an upper story without having the working on the anvils jar the building, has been a problem that has frequently given manufacturers trouble. A mechanical engineer says it may be safely done by placing a good heavy foundation of sheet lead on the floor, and on that putting a good thickness of rubber belting.

Another person who is interested in the problem has tried the experiment, with some success, of placing the block, not on the floor, but on the joist direct, making a cement floor up to the block, and over the wooden floor, reaching back beyond the reach of sparks. It is sometimes said that blacksmith shops never burn, but they keep right on burning in spite of theory, and cement floors ought to be helpful in guarding against fires.

STEAM AS A CLEANSING AGENT

For cleaning greasy machinery nothing can be found that is more useful than steam. A steam hose attached to the boiler can be made to do better work in a few minutes than any one is able to do in hours of close application. The principal advantages of steam are, that it will penetrate where an instrument will not enter, and where anything else would be ineffectual to accomplish the desired result. Journal boxes with oil cellars will get filthy in time, and are difficult to clean in the ordinary way; but, if they can be removed, or are in a favorable place, so that steam can be used, it is a veritable play work to rid them of any adhering substance. What is especially satisfactory in the use of steam, is that it does not add to the filth. Water and oil spread the foul matter, and thus make an additional amount of work.

MIXTURE FOR CLEANSING RUSTY STEEL

Tin putty, 10 parts; prepared buckshorn, 8 parts; spirits of wine, 25 parts. Mix to a paste. Rub on the part to be cleaned and wipe off with blotting paper.

HOW TO CUT A GLASS GAUGE TUBE

Take a three-cornered file and wet it, hold tube in left hand with thumb and index finger at the place where you wish to cut, saw it quickly two or three times with the edge of the file; then take tube in both hands, both thumbs being on the opposite side to the mark and about an inch apart, then try to bend the glass, using the thumbs as fulcrums.

Too much bearing surface in a journal is sometimes worse than too little.

Steel hardened in water loses in strength—but hardening in oil increases its strength, and adds to its toughness.

RULE for roughly figuring on the **coal in a bin or box**—Multiply the length of the bin or box with its width and the product by the height of the coal in feet. Multiply the result by 54 for fine anthracite coal or by 50 for bituminous. The answer will be in pounds. Divide by 2,000 to get tons.

If a **leather belt** is oil-soaked, sift Fuller's earth (a mixture of clay and silicious matter) or prepared chalk on its face, and after a while remove it by scraping with a sharp edged stick.

A little damp salt applied to the pulley side of a leather belt roughens it and prevents slipping.

Oil is injurious to **rubber belts**, but when a rubber belt slips on account of dust and dryness, a little boiled linseed oil lightly applied on the pulley side of the belt will remove the trouble.

ELECTRICAL MACHINERY

ELECTRICITY

Q.—What is electricity?

A.—Electricity is the name for the cause of a large and important class of phenomena in nature, such as attraction and repulsion, heating, luminous and magnetic effects, chemical decomposition, etc.

Q.—Is it a fluid?

A.—It probably is not. Nobody knows exactly what it is. It is now supposed to be a quality, possessed to some degree by all or most substances, consisting in a peculiar movement or arrangement of the molecules.

Q.—Is electricity a newly discovered power?

A.—Its most simple effects were noticed by a Greek, Thales, in the sixth century before Christ. He observed that amber, when rubbed with silk, attracted light bodies, like bits of bran, cork and the like. (The Greek for amber is *electron*, hence the term electricity.)

Q.—What is meant by positive and negative electricity?

A.—It is found that glass rubbed with silk attracts, while the silk repels. Sealing wax rubbed with silk repels, while the silk attracts.

The vitreous (glass) electricity is called positive, the resinous (wax) electricity is called negative.

Q.—Is electricity always due to friction?

A.—No, we have *frictional* (or *statical*) and *voltaic* (or *current*) electricity. The statical is so called because it is *at rest*.

Q.—Which of the two is used in arts and mechanics?

A.—The voltaic.

Q.—How is it produced?

A.—Either by a voltaic battery, or by revolving a coil of wire in the *magnetic field* between the poles of a steel magnet (electro-magnet), or by *inducing* the current by the action of another current or magnet.

Q.—What is induction?

Q.—The process of creating electric properties in a body through the influence of a neighbouring body, having those properties.

Q.—What is an electro-magnetic field?

A.—The space traversed by the lines of magnetic force produced by an electro-magnet.

Q.—What is the principal difference between statical and current electricity?

A.—Current electricity has little electro-motive force, but is very large in quantity. It has little power to overcome resistance (of a non-conductor), but it can do a great amount of work. Statical electricity has the opposite qualities.

Q.—What causes lightning?

A.—It is supposed that lightning is due to the high potential created by the union of many minute water-drops into larger ones and the accompanying immense decrease of surface. What produces the atmospheric electricity, is unknown.

Q.—What is current electricity mostly used for?

A.—For producing the electric light, for electro-plating and for the transmission of energy.

Q.—Is this manner of transmission of energy inexpensive?

A.—Yes. It may be transmitted over miles of wire, which could be done in no other way, and some dynamos transform as high as 90 per cent of the mechanical energy used in revolving the armature into the energy of the electric current.

Q.—What is chemical and thermal electricity?

A.—Chemical electricity is produced by chemical action; thermal is produced by the application of heat to an arrangement of metallic plates.

Q.—Does electricity pass through all substances?

A.—No. Some materials, like rubber, mica and fiber, offer such high resistance that the current will take some other path. They are called non-conductors and are used for insulating conductors.

Q.—What are the principal subjects considered under the head of current electricity?

A.—They are the effects of the current in causing chemical decomposition in electrolysis and electro-metallurgy; in producing heat and light in a resisting medium; in the production of induced currents in a coil of wire; the measurement of electro-motive force (unit: one volt), of resistance (unit: one ohm), of the force of a current (unit: one ampere), and of working power (unit: one watt).

Q.—How many kinds of current are distinguished?

A.—Three—continuous, alternating and multiphase. The continuous current is a constant flow from the positive pole, as in chemical electricity. The alternating current is produced by a rotation of the two legs of a magnet opposite an armature, or, in modern machines, by a rotation of an armature between the two poles of a magnet. The multiphase current results from combining 3 or more alternating currents, with phases displaced with respect to each other; has definite direction of flow.

Q.—Can you detect the nature of a current?

A.—Yes. A magnetic needle introduced into a continuous current will assume a fixed position; in an alternating current it will swing from side to side; and with a multiphase it will revolve.

Q.—How can it be determined which is the north or positive pole in an electro-magnet?

A.—According to "Ampere's Rule," the experimenter considers himself to be swimming head foremost *with* the current, along the wire, always *facing* the iron core; then the **north-seeking pole** will always be at his **left hand**.

Q.—What other way is the positive or negative pole found?

A.—First saturate a piece of white blotting paper with a solution of potassium iodide diluted in a glass of water, parts 1 to 4. Place the blotting paper on one of the brushes, then hook one end of a piece of insulated wire, the ends of which are bare, to the switch and place the other end on the blotter where it touches the brush, and if it is the positive pole or brush the blotter on the under side will turn a brownish color; if it is the negative or south pole it will not affect it.

ELECTRICAL TERMS

Accumulator, or Secondary Battery—An apparatus for storing electrical energy produced by another apparatus.

Alternate Current Dynamo—A dynamo in which the current rapidly alternates or reverses its direction from positive to negative.

Ammeter, or Ampere Meter—An instrument for measuring the rate at which a current passes through a conductor.

Ampere—The unit by which the flow of current is measured—so called after Ampere, a French scientist.

Ampere Hour—A current of one ampere flowing for one hour. When multiplied by the pressure in volts it gives the consumption of electrical energy in Watt-hours, 1,000 of which form the B. T. U. (Kilowatt).

Ampere's Rule—for finding the direction of a current—A magnetic needle, if placed near a current of electricity flowing from the observer who is facing the needle, is deflected to his left.

Anode—The positive terminal of an electric source, in opposition to Kathode, the negative terminal.

Arc—The bow of light produced by the electric current flowing between two carbon points (electrodes) which are slightly separated.

Arc Lamp—A device for regulating and feeding the carbons of an electric arc, so that as the carbons are consumed the distance between them or the length of the arc is continually preserved.

Armature—That portion of a dynamo which revolves between the magnets and in which the electric currents are induced.

Automobile—Machines that move automatically through electricity or any other force.

Bare Conductors—Electric wires or conductors with no covering or insulation.

Batteries, Primary—A set of cells for generating electric currents by chemical action.

ELECTRICAL TERMS 255

Bitumen Insulation—A prepared bitumen compound used for covering or insulating electric conductors.

Board of Trade Unit (B. T. U.) —A measurement of electrical energy decided upon by the Board of Trade for the public supply companies to base their charges upon. It is equal to 1,000 Watt-hours, or about the amount of electrical energy consumed by seventeen 16-candle-power lamps burning for one hour.

Brush of Dynamo—An arrangement of copper wires, gauze or strips soldered together at one end, for collecting the current from the commutator of a dynamo.

Buckling—A bending and displacement of the plates of an accumulator, caused usually by discharging the current too rapidly.

Cables, Electric—Usually applied to electric conductors, consisting of stranded wires, to distinguish them from single wires.

Calibration—Standardizing or correcting of any instrument to the standard value, such as voltmeter, ammeter, etc.

ARC LAMP

Candle, The Standard—A spermaceti wax candle, burning 120 grains per hour, taken as the standard of reference for measuring the luminosity, or candle power of any light.

Carbons—For arc lamps, rods, or pencils, generally made from powdered gas-coke, hardened into shape by baking, and used for the electric arc.

Casing, Wood—A covering or sheath of wood, generally containing two grooves, used for the protection of insulated wires.

Cathode—The negative terminal of an electric source. See Anode.

Cell—A box or other receptacle containing the elements and solutions necessary for the production of storage of electrical energy. A number of such cells are termed a battery.

Change Over Switch—A switch for changing electrical connections from one source of supply to another.

Charging—Filling or storing an accumulator with electrical energy.

Circuit—A system of metallic or other conducting bodies placed in continuous contact and capable of conveying an electric current.

Commutator—Bars of copper and sheets of isinglass to separate them, which form the ends of the armature coils, and from which the current is collected.

Conductivity—The facility offered to the passage of electric currents through a substance.

Conductor—A substance through which electricity will pass, but applied principally to those in which very little resistance is offered to the passage of a current, such as copper wire.

Continuous Current—A current from a dynamo or battery which does not vary in direction and flows continuously.

Controller—An automatic magnetic regulator for a dynamo-electric machine.

Converter — The inverted transformer or induction coil, used on alternating current systems.

Coulomb—The unit of electrical quantity. That quantity of electricity which would pass in one second through a resistance of one ohm with a pressure of one volt.

Current, Electric — The flow of electricity through any conductor.

Creeping—A leakage of electricity over the surface of an insulating body, caused by a film of moisture and dirt, or deposit from evaporation, forming a conductor.

Dielectric—Another term for insulator.

Diaphragm—A plate or sheet securely fixed at its edges, as a drum head, and capable of being set in vibration, like a telephone diaphragm.

Dimmer—A choking coil employed on transformer circuits to regulate the potential. Used in theaters to turn the lights up or down.

Distributing Board—A board from which branch wires or cables are led to various positions from main conductor.

Dynamo—A machine for producing electricity by transforming mechanical work into electrical energy.

Earth or Ground—Term used to denote the leakage of electricity (short circuit).

Earth Return—A circuit in which the earth forms part of the conducting path. It is usually formed by connecting the ends of an insulated line, either to gas or water pipes, or to metal plates buried in the earth.

Electric Motor—A machine similar to a dynamo, but used for conveying electrical energy into mechanical power.

Electrical Energy—The capacity of electricity for doing work, whether for electric lighting or for power or traction purposes. It is directly proportionate to the amount of current and its pressure. Thus by multiplying the flow of current in amperes by the pressure in volts the amount of electrical energy is obtained in watts.

Electricity, Thermal—Produced by the application of heat to an arrangement of metal bodies.

Electrodes—The two terminals forming the positive and negative poles in a battery.

Electrolier—A device for suspending a group of incandescent lamps; the equivalent of chandelier, gasalier, etc.

Electrolysis—The process of chemically separating the component parts of any substance by means of electricity.

Electrolyte—Any substance capable of undergoing a chemical dissolution by an electric current.

Electro-Magnet—A bar of soft iron temporarily magnetized by the influence of an electric current passing through a wire encircling it.

Electro-Metallurgy—The science or process of electrically decomposing solutions or salts of metals.

Electro-Motive Force (usually written E. M. F.) —The cause of the transfer of electricity, and therefore the force which supplies the pressure to an electric current.

Electro-Plating—The depositing of metals by means of electricity upon the surface of another metal or other substance.

Field, Electro-Magnetic—The space traversed by the lines of magnetic force produced by an electro-magnet.

Filament of an Incandescent Lamp—The thread-like substance composed usually of vegetable matter (such as bamboo, cotton, paper, etc.), which by the application of intense heat has been carbonized.

Forming Plates—The operation of bringing the plates of accumulators into proper chemical condition.

Galvanic Electricity—Produced by chemical action; so termed after Galvani.

Galvanometer—An instrument used in testing, for showing the flow of an electric current.

Glow—A white, bright heat.

Henry—The practical unit of self-induction. A secohm or quadrant.

Horse-Power—To find the power of engine required to run a dynamo, multiply voltage by amperes, then multiply the answer by number of lights lit and divide by 746. Answer in Watts.

Hour Lamp—A service of electric current which will maintain one electric lamp one hour.

Incandescent Lamp—A glass bulb or globe from which the air has been exhausted, containing a carbonized filament which comes to a white glow on the passage of an electric current.

Induced Current—Electricity produced by the influence that one magnetic or electrified body has on another not in contact with it.

Induction—The influence that one magnetic or electrified body has over another produced by a dynamo.

Installation—Plant.

Insulation — The non-conducting substance applied to the surface of an electrical conductor to prevent leakage.

Insulator—Any non-conducting material, such as gutta-percha, india-rubber, china, glass, okonite, etc.

Jablochkoff — The inventor of the Jablochkoff candle, an arrangement of carbons placed side by side, and separated by a suitable non-conducting substance, such as kaolin, and used to form an electric arc.

Kaolin—The finest of china clay.

Kathode—See Cathode.

Kilowatt—1,000 Watts.

Mains — Copper cables or other means used for the purpose of conveying electricity, chiefly applied to the larger conductors or cables.

INCANDESCENT LAMP AND SWITCH SOCKET.

Megohm—A unit of resistance; equal to one million ohms.

Meter, Electric—An instrument for measuring the amount of electrical energy used.

Milliampere—The one-thousandth part of an ampere.

Motor—Any machine which may be used for imparting mechanical power. A dynamo running the reverse way.

Negative—See Positive.

Non-Conductor—Any substance which resists the passage of electricity, chiefly applied to those in which this quality is strongly marked.

Ohm—The unit by which the resistance offered to the passage of an electric current is measured; the legal ohm is the resistance offered by a column of pure mercury, 106 centimeters in length and 1 millimeter square in cross-section; or the resistance offered by a copper wire 32 gauge, 10 ft. long (from Dr. G. S. Ohm).

Okonite—Composition of tape and rubber mixed, to wrap joints, to insulate, etc.

Parallel Wiring—Term used to express the system of electrical distribution, in which each lamp has its individual flow and return wires, no current passing through two lamps in series.

Permanent Magnet — A piece of steel or loadstone containing enduring magnetic force and requiring no electric current to magnetize it.

Photometer—An instrument for measuring the intensity of light.

Pilot Lamp—A test lamp frequently used in th

engine-room, serving to denote the E. M. F. of the current from the dynamo.

Plugs, Safety-Fuse—The movable portion of the safety-fuse, containing the fusible wire.

Plugs, Shoe—The movable portion of a shoe or small attachment, to which are attached the flexible wires in connection with the portable lamp.

Poles—General term to express the positive and negative conductors in electricity, or the north and south extremities of a magnet.

Positive and Negative—Terms used to distinguish the polarity of wires in an electric circuit; the flow is termed the positive pole, and the return the negative.

Potential—As heat tends to equalize between two bodies of different temperature, so electricity tends to equalize between two points of different potential. As the difference in level between two water reservoirs connected by a pipe determines the velocity of the equalizing process, so the difference of potential determines the electromotive force of the equalizing electric current. Another determining part is the resistance of the connecting conductor; as the diameter of the connecting pipe and friction are for the two water reservoirs.

Primary Cables and Wires—In an electrical system of distribution where high pressure current is transformed to low pressure, all cables and devices conveying the high-pressure current are termed primary.

Resistance—The opposition afforded by any substance to the passage of electricity.

Resistance Coil—A coil of wire used for creating a certain desired resistance to the passage of a current.

Rheostat—An instrument consisting of one or more resistance coils for varying the resistance in an electrical circuit.

Rocker—An attachment on the bearing of a dynamo to permit of the adjusting of the brushes.

Safety Fuse, or Cut-Out—A device for automatically stopping the flow of electricity in case of accidents or defects in the conductors; a single-pole safety fuse controls only one wire, a double-pole controls both the positive and negative.

Scaling in Accumulators—The formation of a deposit upon the plates which prevents the acid from acting upon them.

Secondary Wires—The low-pressure coils in a transformer, which are acted upon by the primary or high-pressure wires.

Series, Electro-motive—An arrangement of the metals, so that each is positive with reference to those which follow in the list, and negative to those which precede. In dilute sulphuric acid the order is zinc, lead, iron, copper, silver, platinum, carbon.

ELECTRICAL TERMS   265

Series Wiring—Where the positive pole of each cell is connected to the negative pole of the next cell. In the multiple arc, all the positive poles are wired to one post and all the negative ones to another.

Short Circuit—A term used to express any metallic or other connection formed accidentally between a positive and negative wire, by which the current may take a short cut, instead of completing its journey through the lamp, motor, etc.

Sunbeam-Lamps—Incandescent lamps of high candle power.

Switch—An arrangement for breaking or completing an electric circuit.

Telpherage—A system of overhead transportation of goods by means of cars running between two steel rails top and bottom of car from which an electric current is obtained to work motors fixed on one or more of the cars.

Tension—The same for electricity, as pressure for steam.

Terminal—Attachment screw, by which a current enters or leaves any electrical apparatus or conductor.

Thermo Pile—A combination of certain metals coupled together so as to produce electricity by the application of heat.

Three-Wire System—A system of distribution in which two dynamos and three wires are so connected that the third wire serves as flow and return to the other two wires. Besides a considerable saving in the cost of the cables, a constant potential service results.

Transformer — An instrument for reducing or transforming a high pressure current to a low one, or the reverse.

Transmission of Power — The operation of conveying or transmitting power from one point to another.

Turbine — A machine for utilizing the force or fall of running water.

Two or Three-Way Switch — A switch having two or three contact pieces attached to conductors, which by means of a movable handle permits the current to be sent into either conductor.

Unit, Board of Trade. See B. T. U.
Unit, of Current. One Ampere.
Unit, of Electrical Energy. One Watt.
Unit, of Pressure. One Volt.
Unit, of Resistance. One Ohm.

Volt — The unit by which the electro-motive force or pressure of current is measured. It is the E. M. F. that will cause a current of one ampere to flow against a resistance of one Ohm. The volt is based on the product of one Daniell cell. Named after Volta, an Italian scientist and inventor of the Voltaic column.."

Volt-Meter — The instrument for measuring the pressure or E. M. F. of a current.

Vulcanized India Rubber — India rubber, treated with sulphur, etc., to preserve and make it hard. To combine india rubber with sulphur by heat.

Watt, The — The unit by which electrical work is measured. It is equal to the current of one ampere flowing at a pressure of one volt. The amount of energy is found by multiplying the amount of current by its voltage pressure. For instance, a current of 10 amperes with a pressure of 100 volts, represents 1,000 Watts. See B. T. U.

Wire, Flexible — A conductor composed of a large number of fine wires stranded together, so making it flexible.

Wires, Electric — Small conductors, other than the mains.

THE DYNAMO AND ITS PARTS AND ATTACHMENTS

Q. — What is a dynamo?

A. — The dynamo, or better, the dynamo-electric machine, *converts energy* (motion of piston and disc) *into electricity* by the aid of the permanent magnetism present in certain iron portions. The electricity generated then reacts on the iron, heightening its magnetism; the increased magnetism again produces more powerful electrical effects, and so on, until a *limit* is reached. The limit depends partly on the velocity of motion partly on the quality and proportions of the iron and wire in the dynamo, and partly on the resistance throughout the circuit.

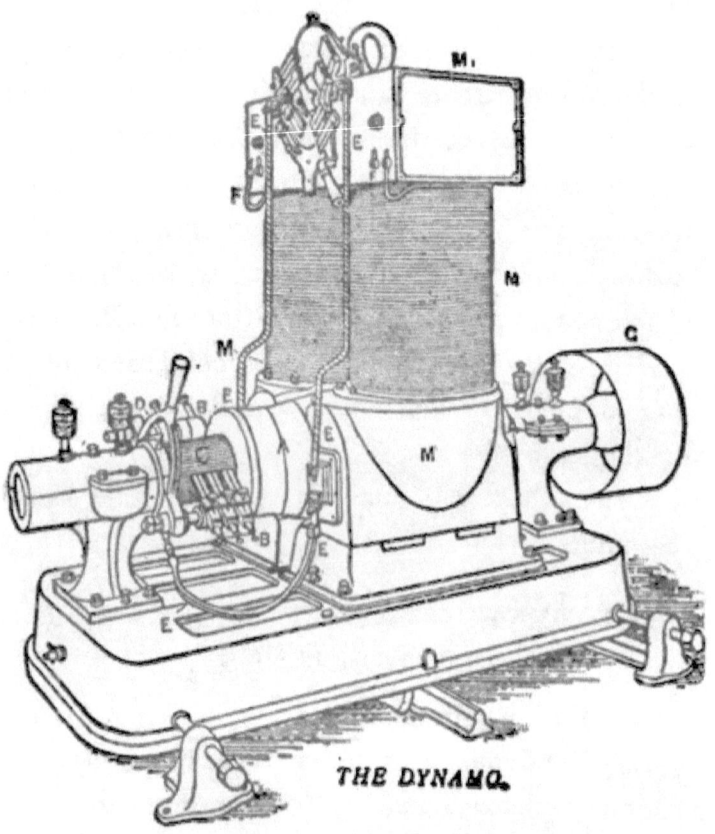

THE DYNAMO.

Q.—What are the parts of a dynamo?

A.—An electro-magnet M, M, M, which is made of two columns of soft iron, encircled by coils of insulated copper wire, and which are united together by cross pieces top and bottom.

THE DYNAMO AND ITS PARTS 269

Between the poles or magnet revolves the armature A, which consists of a number of coils of insulated wire wound around an iron core.

The ends of each coil are connected to copper strips (segments, or bars) placed side by side, forming a cylinder known as the commutator C, from which the current is collected. Generally two sets of so-called brushes or collectors B are fixed upon the rocker (yoke) D, which remains stationary, unless it is necessary to adjust the position of the brushes around on the commutator. (See pages 273, 285.) Attached to these brushes, one set being positive and the other negative, are cables E, E, conveying the two main currents (positive and negative) generated to the switch at the top (side) of dynamo, by means of which connection can be made with the main supply cables. An attachment F, F to convey the current to the electro-magnet is at the top of the two magnet coils. G is the driving pulley.

The armature should be kept up to the proper speed found stamped on field plate. The speed is known by a speed indicator and timepiece. Commutator should be kept quite clean and bright by wiping it occasionally with a rag when running. If necessary it may be cleaned with sandpaper before starting for regular run, the brushes being raised off the commutator.

The brushes for service must be set firmly in

their holders and rest well on the commutator so as to make good contact. They must be set exactly opposite each other and no brush wires (if made so) left straggling.

The rocker yoke holding the brushes should be moved up or down, so as to adjust them to the neutral point, according to the amount of current the dynamo is supplying. When properly adjusted there should be no sparking.

Q.—What is a commutator?

A.—It consists of a number of metal cylinder segments insulated from each other by mica.

Q.—What is the exact function of the commutator?

A.—It serves to rectify, or *send in one direction*, the vibrations or *opposing currents* created by the alternate passing of each pole of the armature before the north and south pole of the magnets.

Q.—How is it done?

A.—It is done in different ways. In some systems, springs, sliding over the half cylinders, are so arranged that they always are one in positive, and the other in negative condition. In other systems the armature is rotated so rapidly (1,600 revolutions per minute, and more) that the waves of current succeed each other at such short intervals, that they appear like a steady current, no break in continuity being perceptible to ordinary tests.

Q.—Does this rapid magnetization and demagnetization produce heat?

A.—Yes, overheating of the dynamo is a drawback of this system of commutation.

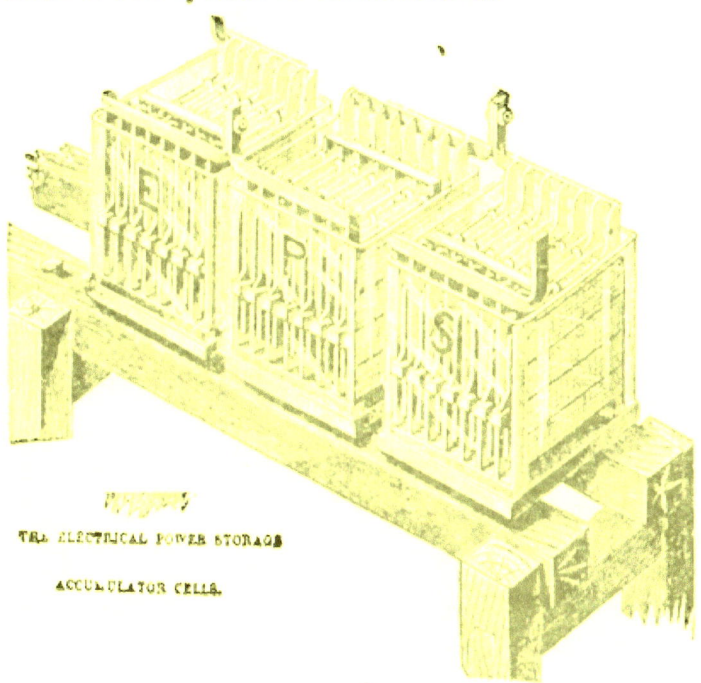

THE ELECTRICAL POWER STORAGE

ACCUMULATOR CELLS.

Q.—What is the accumulator?

A.—It is a battery of cells in which the electrical power is stored. It is also called a secondary battery.

Q.—What is such a battery used for?

A.—The wet battery (large cut) is used for storing electricity to maintain a limited number of lights after the dynamo has been shut down. The dry accumulator (small cut) is used for automobile vehicles, small lamps, etc.

272 QUESTIONS AND ANSWERS

ELECTRIC CYCLE LAMP AND ACCUMULATOR.   POCKET ELECTRIC ACCUMULATOR.   ELECTRIC HAND LAMP ACCUMULATOR.

Q.—What is a rheostat?

A.—It is an instrument for regulating or adjusting a circuit, so that any required degree of resistance may be maintained; a resistance coil.

Q.—Give an example of the way in which the rheostat is used? See cut 🖙

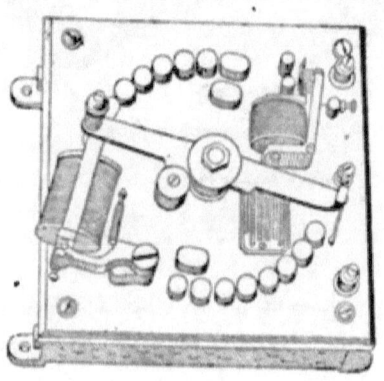

A.—When the dynamo is first speeded to proper speed it shows a dull light and through the rheostat or controller the voltage is raised up to 110, the proper voltage for incandescent lamps.

Q.—What is a transformer?

A.—It is used for tapping a low voltage circuit into a high voltage circuit, as in connecting an incandescent lamp to an arc light circuit.

Q.—What is a "step up" transformer?

THE DYNAMO AND ITS PARTS 273

A.—It is used for the reverse, getting high voltage from a low voltage circuit.

Q.—On what principle are transformers based?

A.—On greater or lesser resistance.

Q.—What is a converter?

A.—The inverted transformer, or induction coil, used on alternating current systems.

Q.—Are the same brushes used for dynamos and motors?

A.—No. The motor brushes are almost exclusively compressed carbon; on the dynamo copper plates or wires are used.

THE TRANSFORMER.

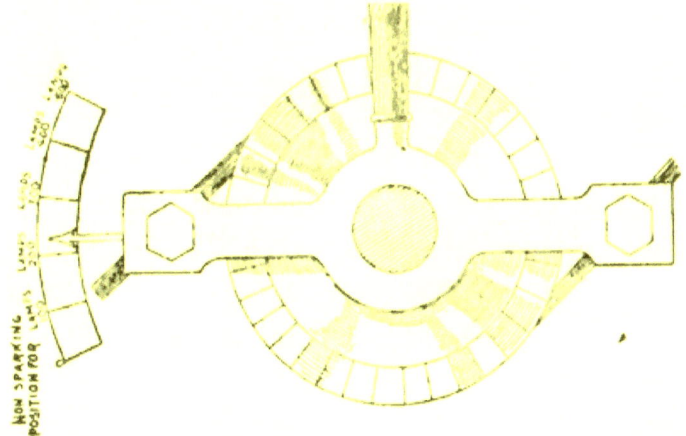

Q.—How are the pointer and scale used?

A.—They are attached to the rocker stud, and serve to secure a sparkless position of the brushes by adjusting the arrow to the scale every time the load (number of lamps in circuit) is changed.

Q.—What is a Daniell cell?

A.—A zinc plate immersed in dilute sulphuric acid contained in a porous vessel, outside of which is a perforated copper plate surrounded by a solution of copper sulphate. The action is as follows: The reaction between the zinc and sulphuric acid produces zinc sulphate and hydrogen. The latter, however, instead of collecting on the copper plate, unites with the copper sulphate, forming sulphuric acid and metallic copper. The former goes to keep up the supply of acid in the inner vessel and the latter is deposited on the copper plate. The consumption of copper sulphate is made good by a supply of crystals in a receptacle at the top.

Q.—What is a gravity cell?

A.—It is a modification of the Daniell cell, in which the porous vessel is done away with. The two liquids are separated by their specific gravities; the copper sulphate surrounds the copper plate at the bottom, and the zinc sulphate surrounds the zinc plate at the top.

HOW TO MAKE TRACING-PAPER

Place your sheets of double-crown tissue paper in one smooth pile, and apply to the top sheet a mixture of mastic varnish and oil of turpentine, equal parts in bulk, using a flat brush, 2 inches broad. Hang each sheet, when coated, over a line to dry. You may trace on this paper with ink.

VARIETIES OF THE DYNAMO

Q.—How would you classify dynamo-electric machines in general?

A.—In generators and motors. A generator is a machine for the conversion of mechanical energy into electrical energy, by means of magneto-electric induction. A motor is a machine for the conversion of electrical energy into mechanical energy by means of magneto-electric induction.

Q.—Why are there so many different generators?

A.—Some are more economical for certain purposes; some give a constant potential, others give a constant current, etc.

Q.—What is a magneto-electric machine?

A.—It is similar to a dynamo, except that the fields are permanent magnets instead of electro-magnets.

Q.—What is a **compound dynamo** machine?

A.—One whose fields are wound with two coils, one of large wire, being in series with the armature, the other of smaller wire in parallel with the armature. This arrangement makes the dynamo self-regulating.

Q.—What is a **multipolar dynamo**?

A.—A bi-polar dynamo has only one pair of field magnets, a multipolar one has more than one pair.

Q.—What is an **alternating current dynamo**?

A.—A dynamo without a commutator. The fields are usually separately excited, as a direct current is required for their excitation.

ELWELL-PARKER ALTERNATE CURRENT DYNAMO.

Q.—What is meant by "separately excited"?

A.—The field coils receive the current for their excitation from some source other than their own armature.

Q.—What is meant by **closed-coil**?

A.—The coils are connected continuously together in a closed circuit, being attached to successive bars of the commutator, as in the Gramme and most direct-current dynamos. When not connected continuously, although attached to successive bars of the commutator, as in the Brush or T. H. arc dynamo, they are termed **open-coil**.

Q.—What is shunt?

A.—A dynamo so constructed that the entire current must pass through the field coils, is called a **series** dynamo; where an additional path (external circuit) is provided, so that only a portion of the current passes through the field coils, this *parallel* connection is called shunt. The fields are "wound in shunt with the outside circuit."

Q.—What is meant by short-shunt?

A.—When the shunt coils of the fields of a compound dynamo are connected to the brushes of the machine, not to the binding posts or external circuit as in the long-shunt.

Q.—What is a "shunt and separately-excited" dynamo?

A.—It is compound-wound, one field coil, receiving current from the armature, the other from a separate source.

Q.—How many kinds of series dynamos are there?

A.—Three, all compound, as follows:

1. Series and magneto, in which the circuit of

a magneto machine is connected in series with its armature and fields.

2. Series and separately-excited, in which the fields have two circuits, one in series with the fields and external circuit, the other being separately excited, used to maintain constant potential at the terminals.

3. Series and shunt, one of the field coils of which is in series with the armature and outside circuit, the other in shunt with the armature.

Q.—What is meant by **synchronizing**?

A.—Modifying the phase of two alternating current dynamos so that they may be connected in parallel.

Q.—What is the three-wire system?

A.—A combination of Edison's for the distribution of electric current for constant potential service, in which three wires are used instead of two, one being a neutral wire. Two dynamos are employed.

Q.—What is **constant potential service?**

A.—An even flow of current. In a water pump the air chamber similarly renders the pressure and flow even or constant. See Potential, p. 263.

Q.—What is the difference between a dynamo and a motor?

A.—The dynamo converts mechanical work into an electric current, which the motor then converts back into, or uses in, mechanical work.

MANAGEMENT AND CARE OF A DYNAMO

THE PLANT

Q.—What rules should be observed in placing a dynamo?

A.—It should be placed in a well-lighted, clean, cool and dry place, and so that it is easily accessible from all sides.

Q.—What should not lie near a dynamo?

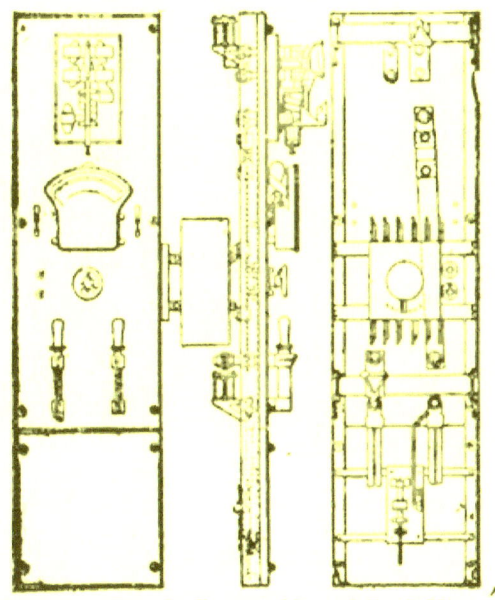

Generator Panel: Front, Side and Back Views.

A.—Iron, steel, bolts, nails, tools of any description, or waste, filings or dust, as they may be attracted by the powerful magnetism or the current of air created by the revolving armature.

Q.—Where should the switch-board and fuse blocks be?

A.—Like the engine, they should be so near the dynamo that the whole plant can be taken

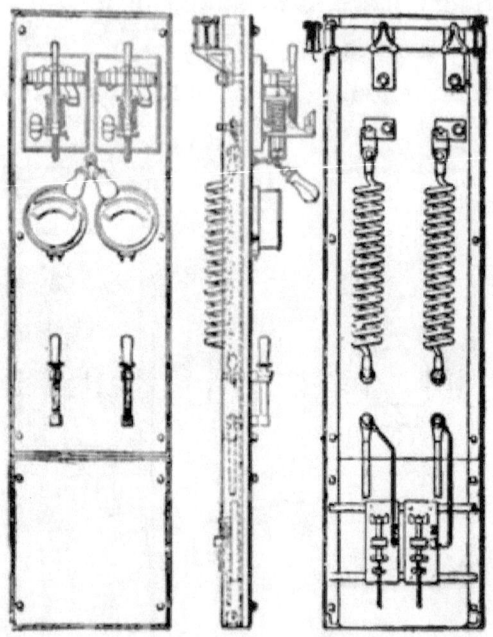

Feeder Panel: Front, Side and Back Views.

in at one glance, but so far apart that there is no danger of a short circuit.

Q.—Of what should the bases of all cut-outs, switches, lightning arrestors, etc., be made?

A.—Marble, slate, or porcelain.

Q.—How should connections be made?

A.—Soldered and thoroughly insulated.

Q.—What kind of wire should be used in damp places?

A.—Rubber-covered wires.

Q.—Would you use metal staples in electric light or power work?

A.—No. Use porcelain insulators.

Q.—What is an insulator?

A.—It is a non-conductor, preventing the current from leaving the wire. (See page 312.)

Q.—In cleat work, what kind would be best to use?

A.—Those with V-shaped grooves; they clamp firmly.

Q.—How many cleats should be used to turn a corner or angle, and why?

A.—Two, to make it neat and workman-like.

Q.—Would it be safe and proper to use a bare wire in any part of the wiring throughout a building?

A.—No. All wires should be properly insulated.

Q.—How do you test the insulation of a wire to see whether it affords the required resistance?

A.—It should be tested to not less than 250 Megohms per mile in dry places and 600 Megohms per mile in damp places. The test must be taken with an electro-motive force of not less than 100 volts after the insulated cables have been in water at 60 degrees Fahrenheit for 24 hours, and with one minute's electrification.

Q.—What should be put in a line, between the cut-out switch and the street where the wire entered the building?

A.—Loops of wire known as drip loops.

Q.—What style of belting should be used for a dynamo or motor?

A.—It should be a light double, endless and rivetless one.

Q.—Why should the belt be endless and not laced?

A.—Because every time the laced joint passed over the dynamo pulley the lights would fluctuate (flicker).

Q.—If the belt is endless, how is the slack taken up?

A.—Nearly all dynamos and motors are provided with a frame and belt-tightening apparatus comprised of one center or two side screws and foundation slides on which the dynamo rests; with these the dynamo or motor can be forced back and the belt tightened.

Q.—How much should a belt be tightened?

A.—Enough to prevent extreme slipping.

STARTING THE DYNAMO

Q.—What should be done every day before starting a dynamo?

A.—The dynamo tender should examine the binding posts, the commutator and brushes, also see that the contacts are clean and firmly tightened by the set screws. Any dust and dirt should be most carefully removed with soft rags and a bellows, as they cause the majority of all the troubles and annoyances.

Q.—What rules should be observed in starting the dynamo?

A.—Always start the machine to running with the main switch open and the brushes raised from the commutator, so all the working parts can be seen; be sure that the rheostat is at zero. Then drop the brushes on the commutator, then see that the voltage is correct on volt-meter. If the brushes spark, rock the brush holder quadrant forward or backward around the commutator until a sparkless place is found, then close the main switch. When running drop a little oil on the end of finger and rub in the palm of hand, then pass the finger gently over the commutator lengthwise.

Q.—How high would you cause the volt-meter needle to rise?

A.—The proper voltage for incandescent lights is 110 volts. If run up to 115, there is danger of burning out the filaments.

Q.—What rules must be observed in disconnecting the dynamo?

A.—In disconnecting the dynamo after it has been used for charging the accumulators or supplying lights, etc., the engine should be eased down, dynamo switch opened, and the brushes raised from the commutator to cool, also to be free in case the armature was turned the reverse way.

The copper brushes should be filed to one bevel, also kept clean and free from oil, copper dust, etc.

RUNNING THE PLANT

Q.—How and when would you test the circuit?

A.—It should be tested every day for grounds, by means of the detector, galvanometer or a magneto bell.

Q.—If there was a ground, how would you locate it?

A.—By disconnecting the circuit in different places, and testing each section separately until located.

Q.—Give various reasons for excessive sparking of commutator and brushes of a dynamo or motor?

A.—Poor condition of the brushes and holders; faulty adjustment of brushes; surface of the commutator rough or covered with dirt and grease; the insulation of one field magnet coil injured and the coil short-circuited in itself.

If one magnet is excited more than the other, one brush will spark more than the opposite one, in the same way as if improperly adjusted.

Two or more segments of commutator short-circuited.

Dynamo or motor overloaded. Overloading will also cause considerable heating of the armature and fields. Overloading of the dynamo, or motor,

THE CARE OF A DYNAMO

may be caused by poor insulation of the external circuit, thus causing a considerable amount of current to escape from one pole to the other.

Grounding of the external wires, which frequently happens in rainy weather.

In arc lighting, lamps may be fed by too strong a current; in incandescent lighting, too many lamps may be put on the leads.

Q.—Name the different causes of the eating away of the segments of a commutator?

A.—Too much tension, too much contact surface, brushes not set properly, or not far enough around on the commutator.

Q.—What will cause flat spots on the face of commutator?

A.—Badly soldered armature wire connections, also soft spots in the copper segments.

Q.—How would you know when brushes have not enough contact or pressure on commutator?

A.—By a peculiar snapping noise.

Q.—What is the effect of a brush being too long, or pressing too hard?

A.—It will cut the commutator, emitting strong spattering sparks.

Q.—How can the brushes be made to press harder or lighter on the commutator?

A.—By adjusting the brush holders.

Q.—How are brushes moved on commutator?

A.—By the yoke (quadrant). (See page 273.)

Q.—Why are they moved around on the commutator, and when?

A.—When more or fewer amperes of electricity are necessary for lights, power, etc.

Q.—How are the different kinds of brushes adjusted?

A.—Practically all alike. They should be set at a bevel of 45° to the commutator. Each brush should cover at least one segment and two insulations to make the current as nearly continuous as possible.

Copper wire or copper leaf brushes are filed to a 45° bevel and the commutator wears them to a concave. Carbon brushes are concaved by putting coarse sandpaper on the commutator, rough side up, and by drawing it to and fro.

Q.—Where should the brushes be set?

A.—At neutral (opposite) points.

Q.—How would you make a copper brush?

A.—Cut the strips the width of the opening in the brush holder, and take so many of them that the brush will pass through the holder easily, and solder them together at one end. The same for wire, gauze and other copper brushes.

Q.—How thick should a carbon brush be?

A.—One-half thicker than the commutator bar to make it strong enough.

Q.—Is it safe to touch the two opposite brushes (positive and negative) at any time while running?

THE CARE OF A DYNAMO

A.—No, for if the motor or dynamo be grounded (short-circuited) the full voltage would be received and cause either paralysis or death.

Q.—When a motor or dynamo becomes very hot, to what would you lay the trouble?

A.—It being overloaded, or poor connections.

Q.—How can stationary motors be reversed?

A.—By changing (crossing) the wires on the yoke or fields, also reversing the brushes.

Q.—Suppose the twine covering the armature wires near the commutator segments happened to unravel while running, what would you do?

A.—Open switch and after motor or dynamo has stopped remove the remaining twine.

Q.—Would it not interfere with the machine?

A.—No. It is there to keep out as much dust as possible from in between the armature wires.

REPAIRS

Q.—What causes the insulation of one or more coils around the armature to char and crumble off?

A.—Excessive heat caused by short circuit, poor connections, overloading, and cotton waste, etc., being attracted at the end of the dynamo and pressed between the armature and pole pieces.

Q.—In what way will the waste injure the armature?

A.—By scaling off the insulation from the wire in some places or bursting the metal bands encircling the armature.

Q.—Do these injuries extend below the outside layer of wire?

A.—Sometimes, but not very often.

Q.—Can the wires be insulated without being taken to a repair shop?

A.—Yes, by carefully lifting one wire at a time just high enough to wrap it with silk tape, and thus insulate it.

Q.—What causes the field magnet to short-circuit and burn the insulation?

A.—By getting parts of the field wire in contact with the iron core.

Q.—What should be done in such trouble?

A.—Unwind the wire until the damaged part is reached, and after insulating it properly, it should be wound back on the cores.

Q.—How are field magnets wound and unwound?

A.—By placing the field horizontally between the two centers of a turning lathe.

Q.—What would you do after having wrapped and insulated all the injured parts of an armature?

A.—Drive the wires back into their positions by means of a hard wood block and hammer, after which give them two or three good coatings of shellac varnish.

Q.—If the injury was below the outside layer, what should be done?

A.—Take the armature to a regular shop and let

them do the repairing, as they have the proper tools to do the work.

Q.—Suppose you had to replace an old commutator with a new one, how would you proceed to do it?

A.—Take the armature out from between the poles, and place the two ends of the shaft on wooden horses. Mark the wires leading from the armature to the commutator by attaching little tags with numbers, to make sure of the proper place of each wire after taking off the commutator. Then disconnect these wires from the corresponding copper bar of the commutator, either by unscrewing the set screw, or unsoldering the connections by means of a hot soldering iron. After this is done remove the commutator, clean the shaft and connections and put the new commutator carefully in its proper position, and connect the armature wires in proper turn to the corresponding copper bars of the commutator by means of set screws, or hard soldering. Great care must be taken not to short-circuit any part of the commutator with drops of solder.

Q.—Why do electrical engineers and linemen wear rubber gloves and rubber soles?

A.—Because rubber, like glass, is a non-conductor of electricity. Live wires should never be touched without one or the other, as instant death may occur.

MEASUREMENTS OF ELECTRICITY

Q.—What is a **dyne**?

A.—The unit of force. The force which, in one second, can impart a velocity of one centimeter per second to a mass of one gramme.

Q.—Why are these terms not given in U. S. measurements, such as ounces and inches?

A.—Because scientists all over the world use the decimal system, and electricity has been developed by scientists exclusively.

Q.—What is an **erg**?

A.—The work done in moving a body through a distance of one centimeter with the force of one dyne. A dyne centimeter. Ten million ergs = one joule. The joule is the practical C. G. S. unit of electrical energy or work. (C. G. S. = centimeter—gramme—second.)

Q.—What is a **watt**?

A.—The unit of electric work or power, equal to one joule per second.

One H. P. = 746 watts.

The number of watts is numerically equal to the product of the current passing, times the voltage which produces that current. 1 volt times 1 ampere = 1 watt; 3 volts times 3 amperes 9 watts, etc.

A kilowatt is 1,000 watts.

Q.—What is a **coulomb**?

A.—The unit of electrical quantity. That quantity of electricity which would pass in one second through a resistance of one ohm with a pressure (force) of one volt.

Q.—What is an **ampere**?

A.—The unit of electric current. That rate of flow which would transmit one coulomb per second.

Q.—What is an ampere-hour?

A.—The equal of one ampere flowing for one hour, or 3,600 coulombs.

Q.—What is an **ohm**?

A.—The practical unit of electrical resistance. A resistance through which an electric current of one ampere will flow under a pressure of one volt.

The *legal ohm*, now internationally adopted, equals the resistance of a column of mercury 106 centimeters in length, having an area of cross-section of one square millimeter, at 0° C., or 32° F.

A section of wire having a resistance equal to the legal ohm is used as a "standard ohm." 1000 feet of $\frac{1}{10}$-inch copper wire has a resistance of very nearly one ohm; a mile of common iron telegraph wire has a resistance of about 13 ohms.

A megohm = one million ohms.

Q.—What is the **Law of Ohm**?

A.—The strength of a continuous current (C)

is directly proportional to the electro-motive force (E) in the circuit, and inversely proportional to the resistance (R) in it. It is, therefore, the e. m. f. divided by the resistance. $C = \frac{E}{R}$; $E = C \times R$; $R = \frac{E}{C}$.

Q.—What is a **volt**?

A.—The practical unit of electric pressure, or electro-motive force. The pressure required to move one ampere against one ohm. The volt is based on the product of one Daniell cell.

Q.—What similarity is there between electrical terms and steam terms?

A.—The volts may be compared to pounds of steam pressure; the resistance to friction; the wire to the pipe; the coulomb to the quantity of steam passing through the pipes; the ampere to the rate at which the steam passes; the watt to the amount of work performed (H. P.).

Q.—What difference do you make between "force" and "power"?

A.—In common language, they are used as equivalents, but in science the following distinction is made:

Force is the cause of a change, such as from rest to motion, or in condition, etc.; power is the rate of the expenditure of energy. Electricity, steam, gravity, expansion, etc., are forces; we speak of horse-power, candle-power, gross and net **power**, etc.

MEASUREMENTS OF ELECTRICITY

Q.—How much of a H. P. is required to maintain a steady light for a 16-candle-power incandescent lamp?

A.—About one-tenth, or 10 lamps to a H. P.

Q.—How many volts for an arc light?

A.—220.

Q.—Can arc lights and incandescent be run on the same circuit?

A.—Yes, by the use of the transformer.

Q.—How is the efficiency of a dynamo determined?

A.—By dividing the electrical energy produced by the mechanical energy expended in driving the dynamo.

Q.—How do you arrive at the amount of electrical power spent?

A.—By multiplying the amount of current by its pressure. For example, a current of 10 amperes with a pressure of 100 volts represents 1,000 watts (1 kilowatt). A current of 20 amperes with a pressure of 50 volts represents the same power (1,000 watts).

Q.—What is meant by the B. T. U.?

A.—It means the consumption of electrical energy of one thousand watts in one hour.

Q.—How are the electric currents measured?

A.—Either chemically or mechanically.

Q.—Explain the chemical system?

A.—It is based on the simple fact that one

ampere of electricity will deposit from sulphate of zinc, under standard conditions, a definite weight of metal. This type of meter is a small electroplating battery through which a certain proportion of the current is carried—the proportion being accurately determined by the relative sizes of the meter wires and the shunts—with the result that one of the two plates is decreased and the other increased in weight, according to the amount of current consumed.

Q.—Where and how is the meter placed?

A.—It should be placed in a clean, dry place and connected to the inside service with moisture-proof wire. It must always be protected by a service cut-out; never placed between the cut-out and the street service. It should be in a place not likely to freeze, also where the inspector can have easy access to it. The meter should be screwed fast to a well-seasoned board, not less than 1 inch thick and well-covered with asphaltum against the wall. Never place more lights on a meter than it is intended to carry.

Q.—What test is made, and how is it made?

A.—It is necessary to find the positive clips and mark them. For this *two things* must be known:

Which side of meter is connected to street, and which of the two outside wires is positive.

The first is found by tracing the conductors or opening the circuit.

The second, by testing with a blotting paper saturated with a solution of potassium iodide. With the moistened paper in hand press it against the upper and middle binding posts.

A brown mark will appear where the paper

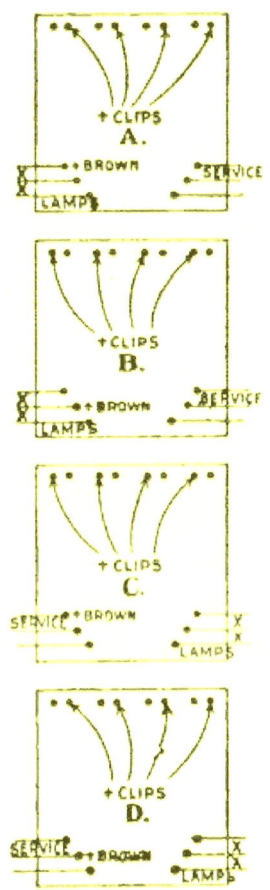

Meter Fed on Right Side.

(See cuts A and B.)

"A"—When top post gives brown mark, right hand clips are positive.

"B"—When middle post gives brown mark, left hand clips are positive.

Meter Fed on Left Side.

(See cuts C and D.)

"C"—When top post gives brown mark, left hand clips are positive.

"D"—When middle post gives brown mark, right hand clips are positive.

touched the positive post. By following rule the positive clips can be determined. The clips are at the top of meter and the chemical bottles set under them.

If the center wire is not brought into the meter, the lowest post must be used for the middle in using foregoing rules.

All positive clips should be carefully marked. The positive plate can be known by it being next to the head of the bolts which bolt the two plates together.

Sometimes the positive plate has a tag attached, which prevents mistakes.

Q.—Explain the mechanical meter?

A.—Of these there are many varieties, those most in use being the Thompson-Houston watt meter and the Westinghouse or Schallenberg ampere meter, both of which are small motors driven faster or slower as the demand for current is greater or less, and communicating their action to a train of wheels with dials like those of a gas meter, so that they may be verified by burning a given number of lamps for an hour and comparing the dials at the beginning and end of the time. The meter record is taken usually once a month, and the bills are based upon these records with as much certainty as though electricity were visible.

THE MOTOR AND CONTROLLER

Q.—What is an electric motor?

A.—The reverse of the dynamo, used for converting electricity into mechanical work.

Q.—When is the work of an electric motor at its maximum?

CONTINUOUS CURRENT ELECTRIC MOTOR.

A.—According to the "law of Jacobi," when the counter E. M. F. is equal to half the E. M. F. expended on the motor, or the impressed E. M. F.

Q.—What are stationary motors used for?

A.—For running elevators, machinery, dynamos, etc.

Q.—Where does a stationary motor get its power from?

A.—From a dynamo, or generally from a three-wire system set of dynamos.

Q.—What voltage, amperes and speed has a 15 H. P. motor?

A.—Generally 220 voltage, and any amount of amperes it may need and 1,600 speed per minute.

Q.—What different systems of using electricity for traction are in use?

A.—The overhead trolley, the underground trolley, the storage battery and third-rail systems.

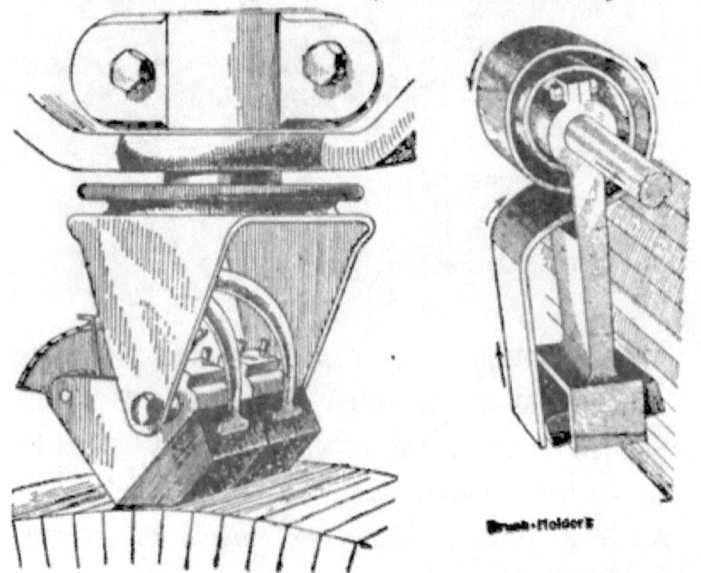

Q.—How are the brushes set on the commutator of a car or engine motor?

A.—Directly against the commutator and opposite, instead of aslant as on stationary motors.

Q.—What voltage and H. P. are the motors of elevated roads?

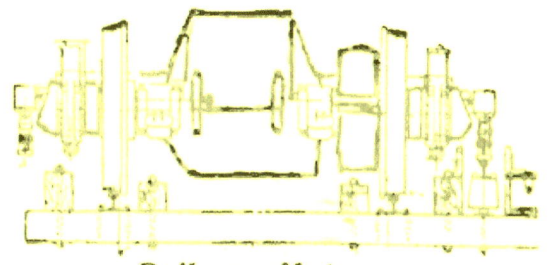

Railway Motor.
THIRD RAIL SYSTEM WITH SLIDING FEEDER SHOE

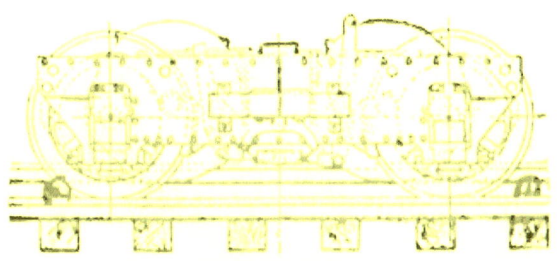

SIDE VIEW
OF
MOTOR TRUCKS

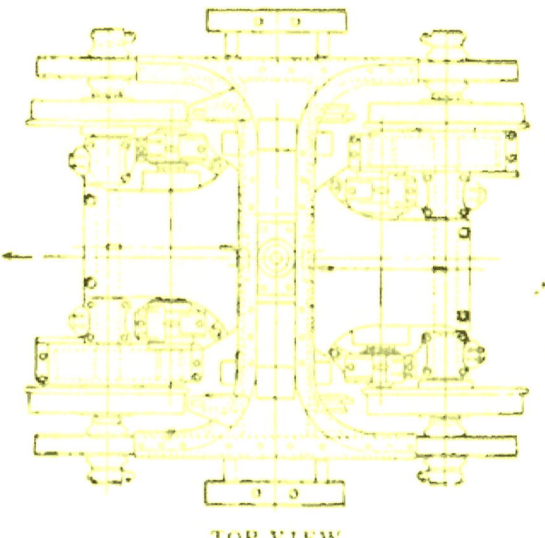

TOP VIEW

A.—Generally 2,000 volt type and 100 H. P.

Q.—How many are usually placed under a car?

A.—Two—one at each end of car, so as to avoid the need of a turn table.

Q.—What system is generally used?

A.—The third-rail system. (See page 299.)

Q.—Can the car be run in both directions from either end?

A.—Yes, by reversing the motor through the controller box.

Q.—How is the electric current transmitted from the third rail to the controller and motor?

A.—By a shoe sliding on the third rail. The shoe may be raised from the rail by a short fulcrum lever in the controller room, thus breaking the circuit.

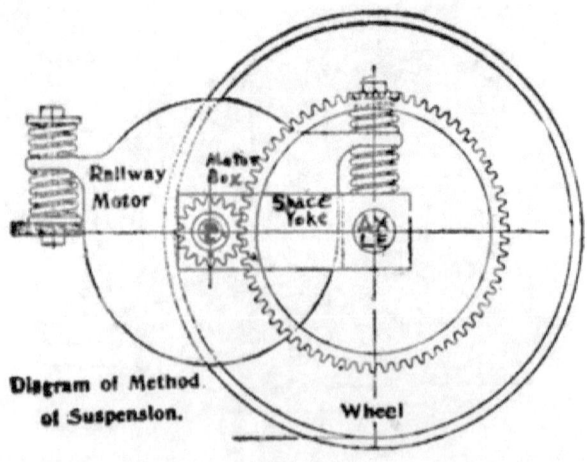

Diagram of Method of Suspension.

Q.—How is the motor of a surface trolley car suspended and geared to the wheels?

A.—It is suspended with springs and a frame between the wheels and truck.

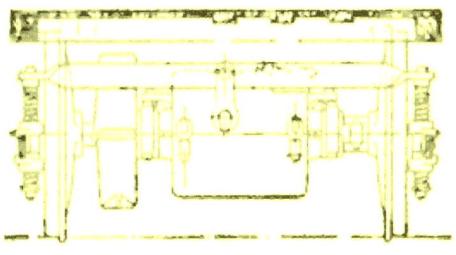

Railway Motor.

The armature axle is geared to the wheel so that the armature pinion turns about four times to a single turn of the wheels.

THE CONTROLLER

Q.—What is a controller and its duty?

A.—It consists of two switches (see cut, page 302), each having its own separate operation to perform. The controlling cylinder (switch) No. 2 proper is used simply to make the different combinations required to obtain the proper regulation of the speed of the car. The second switch, or small handle, is also of a cylindrical form, but is used for either breaking the circuit or reversing the motor, either forward or back.

Q.—Can a controller be compared to a rheostat?

A.—Yes, its function of controlling the speed of the car can be compared to the working of the rheostat in increasing or decreasing the resistance.

Q.—Will switch No. 1 of the controller cut off the current if moved?

A.—Yes, the slightest movement cuts the current entirely off.

Q.—How many notches has the current switch? Name them?

A.—Three—go ahead, back up, center cut-out.

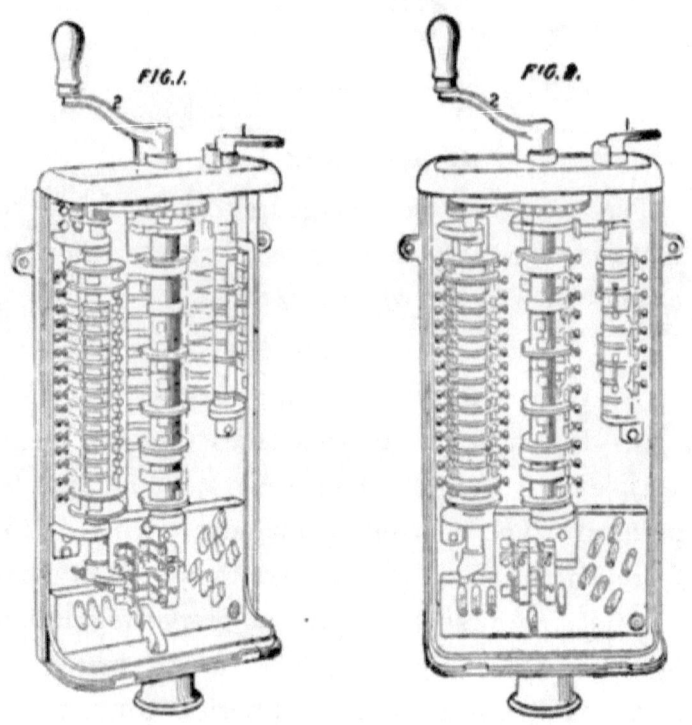

Q.—How is the resistance provided on an electric controller?

A.—A considerable amount of resistance is provided in the shape of bands, fingers or strips of iron or nickel steel.

Q.—How is this resistance subdivided?

THE MOTOR AND CONTROLLER 303

A.—Into a considerable number of parts so that it can be cut down gradually.

Q.—How many notches has a controller dial plate, and what is their purpose?

A.—Generally 7, and they are there to indicate the increase of speed gradually and uniformly by continually lowering the resistance of circuit.

Q.—When the motorman moves the large handle No. 2 around on the dial (notch plate), what does it do?

A.—It connects a coil of wire on the motor with trolley (feeder) wire each notch it is moved.

Q.—Suppose the controller handle No. 2 stood in the seventh notch and the current switch No. 1 was suddenly turned on, what would be the result?

A.—The fuse strips would blow, or melt.

Q.—What are cut-out plugs or strips?

A.—They are fusible wire and are used to save overcharged or heated wires from melting.

Q.—To what could the above be compared so as to be easily understood?

A.—To an attempt at starting a steam engine suddenly at full speed instead of gradually turning on the steam with throttle.

Q.—Compare the principle of controller box and coils of motor with the steam engine?

A.—It is the same as placing the reverse lever in center notch so valve equally covers both steam ports, then opening the steam valve full, this acting as the current switch, and the reverse lever as the controller switch. When ready to start drop the reverse lever or controller down one notch; if more steam or current is required drop it another notch, and so on until the full power or current is passing into cylinder or motor coils (fields).

THE BALDWIN AND WESTINGHOUSE ELECTRIC ENGINE has solved the problem of a locomotive running 120 miles an hour.

In appearance this new wonder does not betray its qualities. The motors are incased, so that hardly any mechanism is in sight. The electric headlight and the pilot alone disclose its character as a motor car or locomotive.

The locomotive weighs 150,000 lbs. and is 37 feet long over the pilot.

The frame is made of 10-inch rolled steel channels, covered by a one-half inch rolled steel plate over the entire floor, giving enormous strength to resist blows in collisions, etc.

This frame is carried on two trucks, with all the modern devices of springs, for swinging motion and free movement. The trucks are built very strong and they are of the swiveling type, so they may go around any curve passable for an ordinary freight car.

The geared connection between the axles and the electric motors permits of any gear ratio desired.

The driving wheels may be connected by parallel rods for pulling heavy trains, as such rods would not permit one pair of wheels to slip without the other.

The motors are directly beneath the car floor, between the two trucks, and are "iron-clad" consequent pole motors.

They are entirely encased in thin steel shells, so as to be protected from all injury under normal conditions of service.

The armatures are laminated, being made up of thin slotted discs of steel. In the slots the armature wires are placed. The commutators are of the best forged copper with mica insulation. The motors have the highest grade of insulation. Power is furnished by the third-rail system.

At both ends is a controller. The path of the current may be divided so as to pass to both motors independently, or it may be sent through one motor to the other.

The braking system has some unique features. The compressed air-brake is used. The engineer's valve is of the standard Westinghouse type. When the handle of the brake valve is put at "emergency" for a sudden stop, pneumatic action breaks the circuit at the same time as it applies the brakes.—A special reversing switch acts on the motors.—The automatic air-pump is driven by electricity, its special motor being directly connected and without gear.

The interior of this locomotive is that of an observation car, and very handsome.

Our 120 miles an hour locomotive is ready for us, but we are not quite ready for it. Before we can risk flying across the country at such speed, all grade crossings must be abolished and the whole present R. R. signal system must be changed. Signals are now about a mile apart, while the new locomotive cannot stop within less than one and a half miles of clear way.

ELECTRICITY FOR HEATING.—To fit heating and cooking utensils for the use of electricity, a thin film of enamel or cement is spread over the outer saucepan, griddle, kettle or heater. Then iron, platinum or other **high resistance** wire is laid zigzag over it, with copper wire connections made to the two ends; and more of the cement or enamel is spread over the wires so as to completely embed them. When enamel is used the apparatus is put in a kiln and burnt on similar to the ordinary iron cooking utensils. In both methods the film of enamel or cement insulating the heating wires is put on so thin and is so good a conductor of heat that the heat generated by the electricity is rapidly conveyed to the utensil

KETTLE.

"SHOE" AND PLUG FOR PORTABLE LAMP.

GRIDDLE.   COFFEE HEATER.

to be heated. Electricity can thus be sent through the wires without fear of overheating them. This would not be possible if they were exposed to the air, which does not conduct heat, but radiates it.

To start motor, quickly throw in switch 4. Then move rheostat-handle 3, *slowly* over the arc, until stopped by magnet 10, which holds it while motor runs. The nearer the handle approaches the magnet, the larger a number of high-resistance wire coils in the rheostat box transmit the current, which has full flow when handle reaches magnet. **To stop motor**, throw switch 4 out, and move handle 3 **off the arc.**

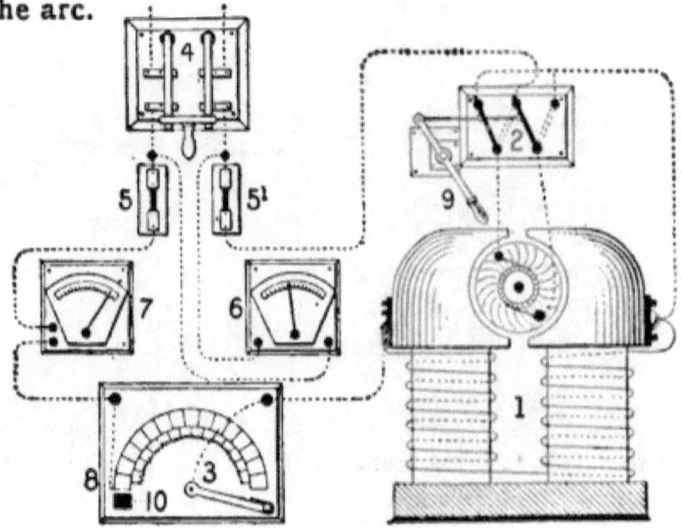

The motor, 1, in the cut is series wound, one pole connected with two binding posts of reversing switch, 2; the other pole connects with handle, 3. 5, 5^1 are safety-fuses, 6 is the volt-meter, 7 the ampere-meter, which at 8 connects with rheostat graduation arc. Fuse 5^1 connects with third binding post on switch, 2. By throwing lever, 9, to the left, the direction of the current, and with it the motion of the motor, is reversed.

ELECTRIC WIRING

CONDUCTIVITY

No part or particle of any substance on earth can ever get lost. It may change its form, but it cannot get away from the earth. There is now exactly as much water on earth as there was one thousand years ago, not even one drop more or less, while there is, of course, a constant change in its distribution over the globe, in the proportions of its three forms (gas, liquid, ice), and in its combinations with other substances. Heat consumed in expansion, etc., is called "latent heat."

So, also, electricity, like any other force, cannot get lost, but it may change its form. When an electric current meets resistance its quantity is *decreased*, and the difference can be indirectly but accurately measured by the increased temperature of the resisting substance. **Current electricity not transmitted is converted into heat.**

The several metals vary as much in their power of transmitting electric currents (conductivity) as they do in other respects. Silver and copper possess the greatest conductivity, tin and lead the smallest. The same fact may be stated thus: Tin and lead offer the greatest resistance to an electric current, silver and copper the least. Any substance that offers great resistance is called a bad conductor; the less the resistance (or, the greater the conductivity), the better a conductor is the substance.

Conductivity is nearly zero for glass, sulphur, resin; it is very low for most liquids and for gases. It varies not only with varying temperature, but also with varying tension, torsion, or pressure. It stands in proportion to the cross-section area of the conductor (wire).

The following table shows at a glance that in the list of the metals named the heat evolved by an electric current increases *in the inverse ratio* as the conductivity decreases. The conductivity of gold is ⅔ of that of copper, the heat evolved is $\frac{3}{2}$ or 1½ times that of copper. The conductivity of tin is ⅙ of that of copper, the heat evolved is 6 times as large:

	Conversion to heat.	Converting power.		Conversion to heat.	Converting power.
Silver	6	to 120	Platinum	30	to 24
Copper	6	to 120	Iron	30	to 24
Gold	9	to 80	Tin	36	to 20
Zinc	18	to 40	Lead	72	to 10

Silver wire conducts 120 units out of 126, losing 6; iron wire conducts 24 out of 54, losing 30; lead conducts 10 out of 82, losing 72.

Silver, copper and gold are excellent for conducting strong currents; platinum and iron are used where very light currents are required, as in telegraphing; lead is used where great resistance is desirable to check too strong a current, as in safety-fuses; platinum is used in incandescent lamps because its expansion in heat is equal to that of glass; zinc and tin find minor employment, as stated further on.

For traction and electric lighting very strong currents are needed, and the conductors are, therefore, made of copper wire of a size proportionate to the rate of flow desired, gold and silver being excluded by their costliness.

Copper can be procured in sufficient quantity, is therefore cheap enough, and is very durable and flexible. The purer copper is the better. The law does not allow an alloy containing less that 96 per cent of pure copper for electrical purposes.

Copper that has become brittle from some cause can be made soft again (annealed) by heating it dark cherry and plunging into cold water.

INSULATION

As dry air is a non-conductor, bare copper wires will conduct a strong electric current without losing any of it (without leakage). This is why telegraph, arc light and trolley wires are left without a covering.

Moist air, all wet substances and water, are excellent conductors, and by one of the principal laws of electricity **an electric current returns to its source by the easiest possible path, or along the line of least resistance.**

The ground, whether earth or lake or sea, is always ready to serve as the easiest and shortest path back to the source, and in order to have the current flow through all the wires of the circuit in undiminished force, and to provide against any portion of it taking a *short circuit*, the wires exposed to possible dampness, contact with water, or the like, are **insulated**, that is, they are covered with a dampness-proof, water-proof, non-conducting material.

Such materials are glass, ebonite, paraffin, shellac, india rubber, gutta percha, sulphur, silk, porcelain, etc. Some of these are suitable in some places and conditions, and others in others.

Iron and porcelain are used for supporting bare wires. For telegraph cables gutta percha is used, which, however, is easily affected by heat, and cannot be used for insulating electric light wires carrying a large voltage. Instead, a fibrous matter (jute and the like), steeped in a resinous or bituminous compound, is used. India rubber (vulcanized to make it harder and more durable) is considered the best insulator for house wiring.

To prevent decomposition of the rubber by the copper, the copper wire is usually co.. which also makes soldering the joints easier. Then a cover of india rubber is put on, then a second cover of vulcanized india rubber. The third covering consists of india-rubber-coated tape (okonite), and over this tarred flax is braided and coated with a preservative compound.

When the wire is so insulated, the electric current finds it easier to travel the length of it for thousands of miles than to escape through the insulation, one-sixteenth or one-eighth of an inch thick. But remember, wherever the insulation is faulty or injured, admitting water or contact with any other good conductor, there the current escapes and returns by the shortest route to the source.

Aside from the loss in lighting power, such defects are frequently the cause of disastrous conflagrations or of death. House wiring should be entrusted to experienced, skillful and conscientious men only.

For test of insulation, see page 281.

SIZE OF WIRE

Two considerations determine the proper size of wire to be used in a circuit: The wire must be thick enough to carry an electric current of the desired voltage at the desired rate of flow, and on the other hand, to avoid unnecessary expense, it must not be thicker than necessary.

If the wire is too thin, a portion of the current is converted into heat (see page 309), and the hot wire becomes a source of great danger. The fire insurance inspectors insist on this point, and rightly so. Of course, the light furnished by too thin a wire is very poor.

CONNECTIONS

A joint or connection must be **solid**, that is, it must not offer any more resistance than the wire itself, and must, therefore, be made with the utmost care. The second requirement is, it must be **damp proof**.

The strands of copper are first cleaned by scraping, then interlapped or scarfed, another wire is wound around the joint (especially in the case of

Joints.

large cables), and the whole is soldered. This gives a so-called *hard joint*. Then insulation is put on with equal care, first india rubber, and then okonite (india rubber tape), making a solid and dampness-proof insulation.

ARRANGEMENT OF CIRCUITS

The plan for the wiring of a building should be given great care and skill, requiring much experience.

First, find the point at which the main circuit from the dynamo or supply-wire will be most conveniently divided up into a number of smaller circuits. A fault is easily located in a small circuit and cannot disturb the service except in its own circuit. A *"central distribution board"* is erected, and from this a small special circuit leads to each lamp, or, in a large building, cables run to a number of branch distribution boards, with which then the lamps are connected.

It is best to connect each lamp with the distribution board by two wires (parallel wiring). If all the lamps are strung along one common circuit (series wiring), all the lamps will be affected by any little irregularity, and the distribution of current is uneven. These disadvantages far outweigh the saving in the first expense of installation.

PLACING THE WIRES

The greatest care should be taken to keep the wires absolutely free from dampness or water, since they establish at once a short circuit (earth, ground, leakage), and a portion or all of the electric current returns by the nearest path (by wall or pipe, or the like, and the ground) to the source.

Another great cause of annoyance to be guarded against, is a short circuit by leakage from one wire to another through defective insulation, crossed wires, etc. The electric current, finding its way through dry dust, lint, rubbish, or wooden parts, heats them to the point of ignition, and a fire is the result.

In placing the wires, neatness of appearance, safety from leakage, and protection from fire are the three points to be kept in mind.

Wires without casings should be 6 inches apart for mains, and 2½ inches for smaller sizes.

Metal or glass tubing is a good protection against gnawing rats or mice, but insurance inspectors do not look upon them with favor for high-voltage circuits.

The most serviceable casings are of well-seasoned hard wood, grooved. The fillets separating the grooves should be 1¾ inches in width for mains, 1 inch for main branches, ½ inch for smaller branches. The inside of wood casings should be painted with a fire-proof paint or compound, and the wires packed in with asbestos or silicate cotton.

For chandeliers twin wires are generally used. They should be handled very carefully, and properly protected with cut-outs.

CUT-OUTS

What the safety-valve is for the boiler, the cut-out or safety-fuse is for the electric circuit. It consists of a short lead or tin wire of a size proportionate to the greatest quantity of current required for the circuit. If the current increases beyond that point, the tin or lead wire, unable to transmit more than so much of the current, converts the excess into heat and melts (is blown), thus breaking the circuit.

The cut-out is a guarantee, therefore, against overloading the wire from any cause, short circuit, crossing wires, or negligence of dynamo attendant.

The best place for the cut-outs is on the distribution board, where the small circuits are connected to the mains. The circuits can thus be easily disconnected by removing the safety-fuse. For special purposes cut-outs are placed wherever desirable.

THE SOCKETS

Incandescent lamps receive the current from contact pieces in the socket of the "shoe," or "plug," or in a socket swinging on the circuit wires.

Such plugs are frequently distributed along the walls of the rooms, for the sake of attaching a portable lamp to either one of them, wherever it may become desirable. This is a great convenience.

The socket is either a screw socket or a "bayonet" socket. The latter is simply pushed in and turned around one-eighth, to make connection with the circuit. In the screw socket connection does not take place until the lamp is screwed in as far as it will go.

To avoid the handling of the lamps when the current is to be turned on or off, a key or switch is provided. This key should never be placed anywhere *between* full-off and full-on, not even for a moment, because in a *partial* connection the contact pieces will get heated and grave consequences may result at once or later on.

THE ELEMENTS OF ALGEBRA

By Prof. O. H. L. Schwetzky

Q.—What is arithmetic?

A.—The science of numbers, or the science of numerical equivalents.

Q.—What does it teach?

A.—It teaches how to calculate or compute quantities by the means of numbers.

Q.—What is algebra?

A.—Sir Isaac Newton called it "universal arithmetic," meaning by this term, that algebra teaches the rules which apply to any and all numbers.

Q.—What is one of the principal differences between arithmetic and algebra?

A.—In arithmetic we have only 10 characters with which to work: 0, 1, 2, 3, 4, 5, 6, 7, 8, 9—and which, besides, have a limited meaning, variable by position only. In algebra, quantities of every kind may be denoted by any characters whatever.

Q.—What are the characters mostly used in algebra?

A.—The known quantities in each case are generally denoted by the first letters of the alphabet, a, b, c, etc., and the unknown quantities to be found are represented by the last letters of the alphabet, z, y, x, w, etc.

Q.—What do these characters represent?

A.—They represent any number chosen.

If we assume a to represent 9, and b to represent 3, then $a + b = 12$; and in $a + b = c$ we would put $c = 12$.

In $a - b = c$ we would have $c = 6$; in $a \times b = c$, c would be $= 27$; in $a \div b = c$, c would be $= 3$.

In $a + b = c + x$, x is the required answer, a, b and c being known quantities.

Q.—What signs are used in arithmetic?

A.—*Plus* (+) for addition, *minus* (—) for subtraction, *times* ($\times$) for multiplication, *by* ($\div$) for division, and *equals* (=) to show equality.

Q.—What signs are used in algebra?

A.— +, — and =, as in arithmetic. The $\times$ is rarely used. Instead of $a \times b$ the form a b is employed, or a . b . Instead of $a \div b$ we write $\frac{a}{b}$.

Q.—Name another difference.

A.—In arithmetic any operation that is readily performed is at once executed and the result substituted, as 10 for $7 + 3$, 3 for $10 - 7$, 21 for 3×7, 7 for $21 \div 3$; but in algebra this is not done: $a + b$ is called a sum; $a - b$ is a quantity equal to the excess of a over b; a b is a product; $\frac{a}{b}$ is a quotient; $(a + b)(c + d)$ is the product of the two sums

$a+b$ and $c+d$; $a(b+c)$ is the product of a and the sum $b+c$; $a\left(\dfrac{b}{c}\right)$ is the product of a and the quotient $\dfrac{b}{c}$, etc.

Q.—Explain the use of the parenthesis, ( ), further?

A.—It means that the term enclosed in the parenthesis is to be treated as **one quantity**. If $a=9$, $b=8$, $c=4$, and $d=3$, then $(a+b)(c+d) = (9+8)(4+3) = 17 \times 7 = 119$; $a(b+c) = 9 \times 12 = 108$; $a\left(\dfrac{b}{c}\right) = 9 \times 2 = 18$.

Q.—Are there no definite numbers used in algebra?

A.—Yes. $a+a$ is written $2a$; $ab+ab+ab = 3ab$, etc. The definite numbers in this case are called numerical coefficients, or for short, *coefficients*.

$a \times a$ is written aa or a^2, which is read "a square." $a \times a \times a$ is written a^3, which is read "a cube," etc. In this case the figure indicates how many times a quantity is to be multiplied by itself, and is called the *exponent*.

Q.—In what relation does a stand to a^2?

A.—It is the square root of a^2.

Q.—What is the meaning of $a - b + \left(\dfrac{c}{d}\right) - (\because ab + d) + 6ac$?

A.—That depends on the definite numbers to be substituted for the characters. If all the $+$ quantities (*positive* quantities), $a + \left(\dfrac{c}{d}\right) + 6ac$ added together give a larger quantity than the — quantities (*negative* quantities), b + 2 ab + d added together, then the answer is positive, otherwise it will be negative.

Q.—How can a quantity be negative?

A.—In the case of bookkeeping it would mean that there is that much deficit or loss; in traveling it would indicate that distance back of a certain point instead of forward; on a thermometer or steam gauge it would indicate so many degrees below zero instead of above, etc. *Plus* means "above zero," or "more than nothing," *minus* means "below zero," or "less than nothing."

Q.—What is the difference between $a - b + c$ and $a - (b + c)$?

A.—In the first case b is to be subtracted from the sum of a and c; in the second case the sum of b and c is to be subtracted from a. The difference becomes clear by substituting definite numbers: $20 - 9 + 4 = 15$; $20 - (9 + 4) = 7$.

Q.—What is the meaning of $a - (b + c) = a - b - c$?

A.—It means that additions and subtractions may be performed in any order. We may either subtract the sum $b + c$ from a, or we may subtract first b from a and then subtract c from the remainder. The result is the same.

Q.—Can you further illustrate the meaning of the *minus* sign?

A.—
1. $a + b = c - b$
$a + 2b = c$

Taking $-b$ away on one side is the same as adding $+b$, because $+b - b = 0$. To keep the two terms at the sides of the $=$ sign of the same value, we must add $+b$ at the other side, too, which gives $2b$.

2. $a(-b) = -ab$

This signifies that "multiplication by a negative quantity" $(-b)$ means "starting from the zero point *in the opposite direction.*" If John has $500 assets, and Frank has ten times as much *liabilities*, he owes $5000. Also: If John owes $500, $(-a)$, and Frank owes ten (b) times as much, he owes $(-a)b = -ab$, or $5000.

3. $(-a)(-b) = +ab$

This is the reverse of the above (2). The same principle applies. If John is $500 short, and Fred has ten times as large an amount of cash on hand, he has $5,000. Expressed as a rule, this simple truth presents itself as follows:

Minus multiplied by *minus* produces *plus*,

or, in other words, the product of 2 negative factors is positive.

Q.—Can you further illustrate this rule?

A.—1. A rich man said to his son: "I will make you the owner of a fortune 8 times as large as your present indebtedness." The son confessed that he owed at that moment $7,000. To make good his promise, the father had to pay the debt and give his son $56,000 besides.

2. One ship sailed 200 miles due east from a port, while another steamed 3 times as far due west. They were consequently 800 miles apart, one being 200, and the other 600 miles, from the port, in opposite directions.

3. An open siphon, one arm of which had a five times larger inside area of cross-section than the other, was provided with a scale, and filled with water to the zero point of the scale. A piston was introduced in the wider arm, and pressed down, until the water surface in this wider arm stood at 3 inches below zero. Where was the surface in the other, smaller arm? Ans. $(-3)(-5) = +15$.

Q.—What is the meaning of $a\left(\dfrac{b}{c}\right) = \dfrac{ab}{c}$?

A.—It means that multiplications and divisions may be performed in any order.

Q.—What advantage does algebra give?

A.—It gives short characters instead of long numbers, and tedious multiplications, etc., are avoided, as no such operations need to be executed, except in the answer, where the given values are substituted.

THE TRACTION ENGINE

A traction engine is a locomotive for common roads, and by throwing the driving wheels out of gear is converted into a stationary engine.

As a **traction engine** it is steered by a worm gearing, which turns a winding shaft, on which a chain is wound and unwound, drawing one or the other front wheel back, according to the direction in which the engine is to run. The engineer steers by turning a hand-wheel controlling the worm gearing.

The driving wheels have V-shaped projections on their rims to prevent slipping. They get their motion through differential or compensating gears from the engine. (See cut page 326.) The motion of the engine is reversed through a special device, a single eccentric reversing gear, or through a reversing rack. (See pages 330, 331).

Coal and water are carried in the combination tank.

For turning the curves of a road one of the drivers is loose on the shaft, so that it may run a longer or shorter distance than the other driver, without straining the axle or connections. For running on a straight road, the loose driver is made solid with the shaft by inserting the key A into slot B.

As a **stationary engine** (the drivers being thrown out of gear), the pulley-face fly-wheel (or a friction clutch wheel, see page 326), furnishes the power by means of a belt.

The rear axles and brackets of all good traction engines are placed back of the firebox, so that the weight will be well distributed between the fore and aft wheels. Short axles riveted to the sides of the firebox are very dangerous.

DIFFERENTIAL OR COMPENSATING GEAR

The differential or compensating gear is arranged as seen in the cut: A is a large bevel wheel (loosely set on the axle), carrying three pinions, B, so distributed over it that they

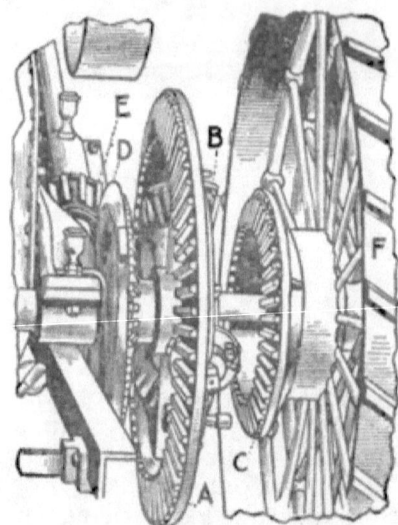

together engage the ground wheel by meshing either with C or D, according as the engine is to travel forward or backward. C is bolted to the main drive-wheel; D is keyed on the axle. A gets its motion from the engine through the inclined shaft and the bevel pinion, E.

FRICTION-CLUTCH FLY-WHEEL

The fly-wheel of a traction engine must allow of being thrown in and out of gear easily. One of the most convenient devices for this purpose is the friction clutch shown in the cut.

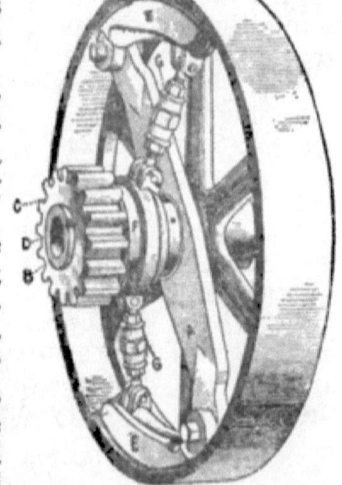

The wheel has diametrically placed a driving-arm, A, to which is cast a sleeve, B, surrounding the axle of the wheel. The end of B carries the pinion C, keyed to it at D. Both ends of the driving arm A have a cast-iron shoe, E, loosely bolted to them. The bolts are solid with A. These shoes are hollow and filled with hard-wood blocks with a surface curved to correspond exactly with the inside surface of the wheel rim, turned true The free ends of the shoes are provided with a toggle-joint (or turn buckle joint), G, G, by which the wood is pressed firmly against the wheel rim, when the fly-wheel is to be engaged. The toggle-joint is worked by throwing the collar F, which loosely fits around the sleeve B, toward the wheel. This is done by means of a lever within easy reach of the engineer, but not shown in the cut.

CROSS-HEAD

A is the piston rod with threaded end. B is the piston lock-nut, C is the cross-head frame. D, D are the slide blocks, E, E are the cap screws which hold the slippers (slides), D, D, to the cross-head frame. F, F are the adjusting screws for taking up the wear of the slides. This is done (about once a year) by slightly slacking out the bolts E, E, and screwing in the screws F, F, until the lost

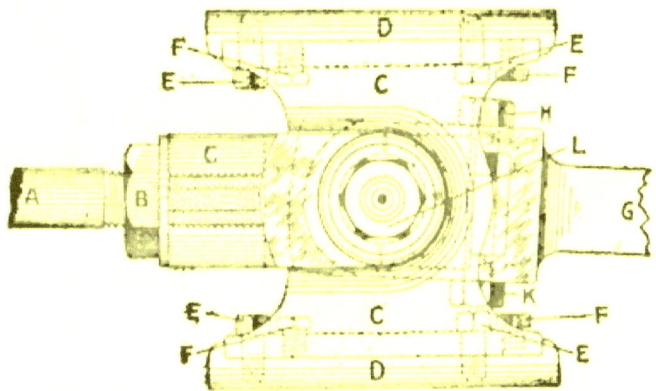

motion is taken up. G is the small end of the connecting rod containing the brasses, adjustable by means of a screw bolt, H, engaging with a beveled block in the strap, said bolt being locked in position by jam-nut K. The cross-head pin, L, can be removed by unscrewing the nut from the pin and driving the pin out of the cross-head frame by means of a hammer and a wooden block.

DIMENSIONS AND HORSE-POWER OF TRACTION ENGINES

Their speed is about 250 revolutions a minute.

Ten-inch Stroke Simple High-Pressure Engine.		Ten-inch-Stroke Compound Tandem.		
Horse Power.	Diameter in inches.	Horse Power.	Diameter in inches.	
			High P. Cyl.	Low P. Cyl.
9	7¼	12	5¾	8¼
12	8¼	15	6⅛	9
15	9	20	7	10
20	10	25	7¾	11

TANDEM COMPOUND CYLINDERS AND VALVE MOTION

The Tandem Compound Traction Engine does not differ much from the plain single cylinder engine, in operation, or in the care it requires. Where the work (load) amounts to the full horse-power capacity of the engine, the tandem compound is economical, otherwise it is wasteful.

The accompanying cut shows the very simple and compact arrangement of the two cylinders (one high and one low pressure) and the slide valve. The cylinders are cast separately and bolted together at U. The partition R is cylinder head for both cylinders, is held in place by jam bolts (Y), and at S the piston rod passes through its center. The packing is metallic and does not need adjustment or renewal. One piston rod carries the two pistons, A, B. By unbolting at U, U, the interior of the two cylinders is reached easily.

Only on the larger or "low pressure" cylinder is there a steam chest, valve seat, etc. In order to connect with both cylinders, the slide valve and seat are arranged as follows: Steam for the boiler enters through H into X, a chamber formed by the hood enclosing the slide valve. This hood and the slide valve proper are one casting, and move together in the steam chest, M. The passages I, I communicate with M. From X the live steam passes through L into E in the high-pressure cylinder. At the end of the stroke, X and K communicate and live steam enters from X through K (see dotted lines) into D at the other end of the high-pressure cylinder, reversing the piston motion. The expanded steam in E exhausts through L into the receiving chamber, M, and from there passes through I and N into G in the low-pressure cylinder. The steam expanded in D exhausts through K into M, and passes from

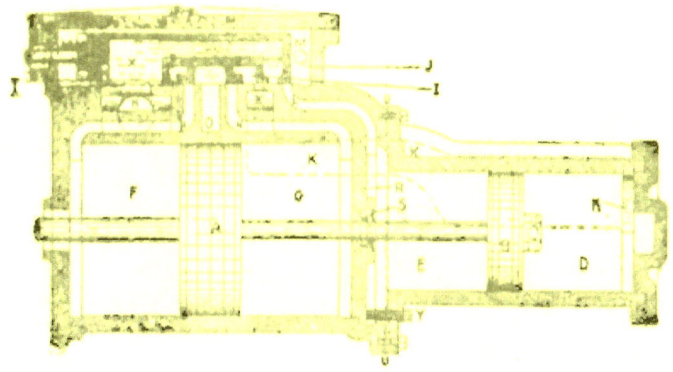

there through I, P into F. P and N finally carry the steam exhausted from the low-pressure cylinder off through O. Port J in M serves to admit live steam to facilitate starting the engine. After starting, it is closed.

The low-pressure cylinder must be larger than the high-pressure cylinder, because the expanded steam, exhausting from the latter into the former, is so much weaker than live steam that it requires more piston area to work on, to furnish the same pressure as the live steam exerts on the smaller piston area. The proportion of the areas is closely calculated by experts; roughly it may be said to be 1:2. See table, page 328.

For more explicit information on compound and other engines see pages 96, 104 and 133.

Some engineers have asked why the valve was not worked directly by the piston rod, by means of a lever of the proper kind and proportion. (See leverage, page 241.) In the beginning the valve was worked in that crude way, and, at the very first, by hand. The necessity of economy, however, in the consumption of steam has led to the devising of eccentrics, link motion, compound engine, single eccentric, etc.

SINGLE ECCENTRIC, VALVE AND ENGINE REVERSING GEAR

For general information about the eccentric, see page 116; link motion, page 144.

A traction engine must necessarily have the most simple possible attachments. A very simple and ingenious valve and reversing gear is shown in the cut.

There is only one eccentric. To the eccentric strap, which carries the valve rod, a roller is pivoted a little above the valve rod pin. The roller runs in a guide, the position of which is regulated by the reverse lever, to which it is connected by a "reach rod." Changing the angle of the guide reverses the engine, or it may simply shorten or lengthen the travel of the valve and thereby change the point at which the steam is "cut off."

Besides its extreme simplicity, this device has the further great advantage that it makes the lead of the valve "constant," that is to say, the lead does not vary with the travel, as it does in link motion. The valve gives a quick, full opening at exactly the right moment, admitting steam promptly at the dead points. Also, it cuts off quickly, giving quick expansion.

To **reverse**, the lever is thrown back to the furthest notch on the quadrant. To **stop**, the lever is placed in the center notch.

By holding the lever between the center notch and one of the other notches, the stroke of the slide valve may be set at any desired length short of its extreme travel. This is done by economical engineers, when the load is light.

The whole gear is so simple and durable that no skill is required to run it, or to replace and adjust it. The eccentric has its center almost directly opposite the crank pin, so that the roller will stand exactly over the center of the guide, when the engine is at either dead center. This renders it easy to find the correct position of the eccentric.

To **set the valve**, the eccentric rod is then adjusted so as to give equal lead at both ends of the valve.

REVERSING RACK

For throwing the eccentric into the reverse position, the reversing rack shown in the accom-

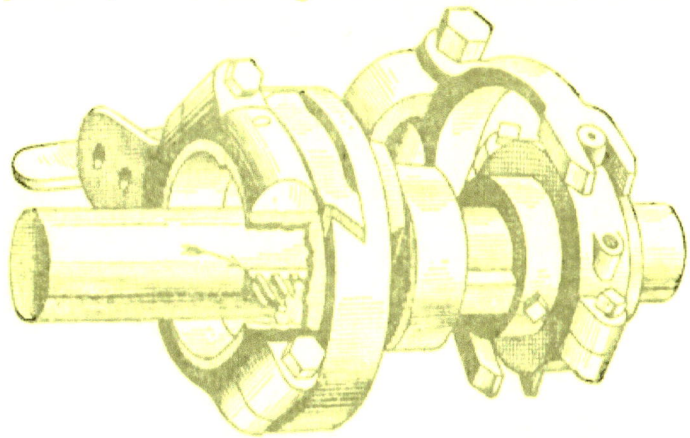

panying cut is used on simple or single cylinder engines. Where the eccentric is cut away in the cut, the arrow shows the reversing rack.

STACKER GEARING AND TURNTABLE

A, A is the upper frame; B, B is the lower frame. C is the lower frame chair bracket, with a central hub, and the pocket D, in which a collar turns, that is attached to the upper frame chair bracket E. The center gear shaft turns in the lower frame hub and in the upper frame collar,

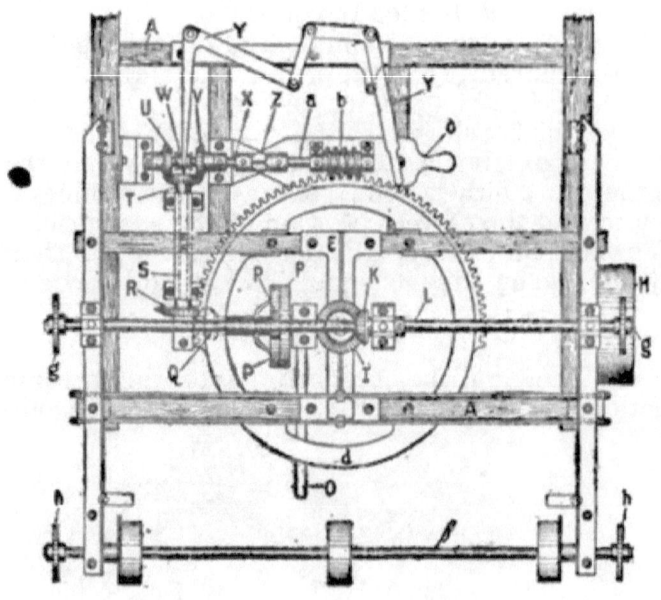

and is geared *below* (miter gear F) to the main shaft, G, driven by pulley H, and *above* (miter gear I, K) to the sprocket wheel shaft L. M is the shaft box. N is the bracket for the pin which, by means of the lever O, serves to clutch the three-wheel gear P, P, P, with the main shaft G, around which it fits loosely. When so engaged the pinion Q at the other end of the sleeve engages pinion R and the shaft S, at the other end of which there is a bevel spur wheel, T, driving U and V in opposite directions. The reversing spool, W, serves to throw shaft X into gear with either U or V. W is worked by means of the lever Y. Z is a universal joint, enabling shaft *a* and worm *b* to be thrown out of gear with the turntable *d* by the lever *c*. The turntable *d* is attached to A, A, pivots around the central shaft in the collar in D, and has a ball-

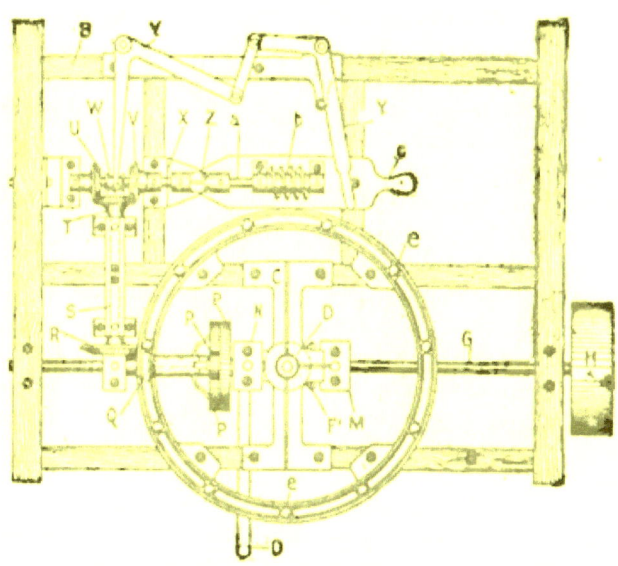

bearing, E, E, attached to the lower frame. The shaft L drives the working shaft *f* by means of the sprocket wheels *g*, *g*, *h*, *h*. The three pulleys on shaft *f* drive the different parts by belting.

Lever O starts or stops the stacker; lever *c* starts or stops the turntable *d;* lever Y controls the direction in which *d* turns. When the turntable is thrown out of gear the stacker may be swung around by hand.

THE STACKER

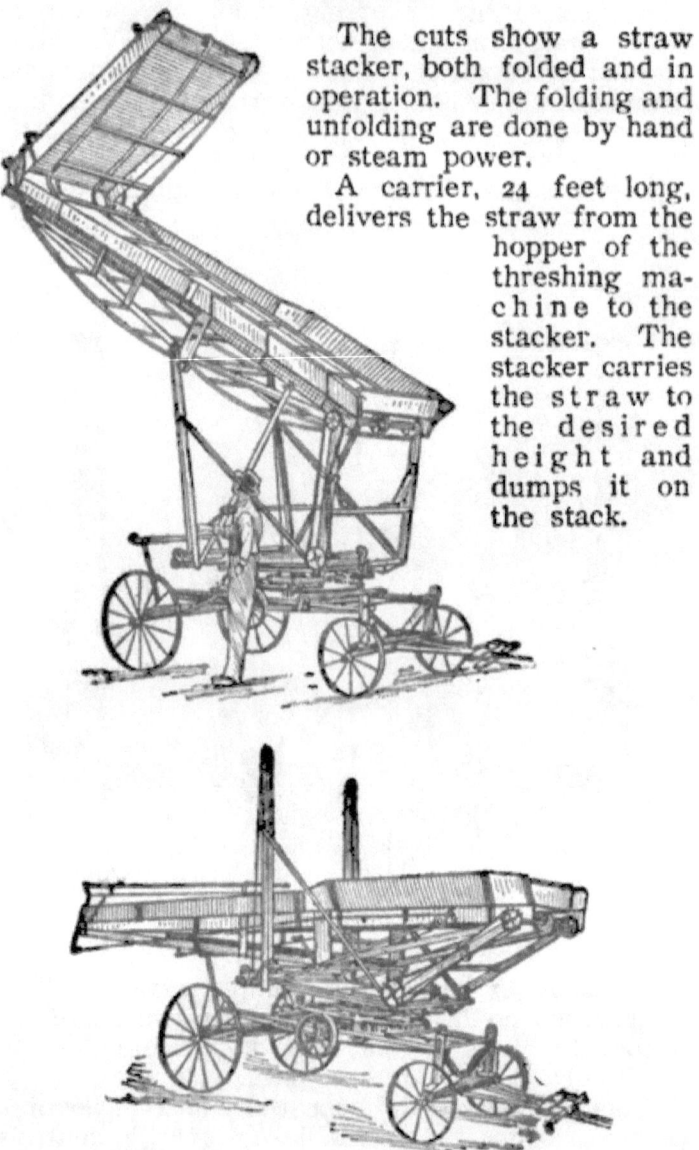

The cuts show a straw stacker, both folded and in operation. The folding and unfolding are done by hand or steam power.

A carrier, 24 feet long, delivers the straw from the hopper of the threshing machine to the stacker. The stacker carries the straw to the desired height and dumps it on the stack.

The mechanism of the turntable, reversing gear, etc., are fully described on pages 332, 333.

JOURNAL-BOX BABBITTING

Standard Babbitts	Martin's Nickel	Copper	Antimony	Tin	Zinc	Total
High Speed Babbitt...	10	16	4	70	......	100
Medium or Common...	......	4	6	90	......	100
Machinery Bearings...	......	88	...	12	......	100
Muntz Metal............	......	60	...	......	40	100
German Silver.........	33½	33⅓	...	......	33⅓	100
White Brass............	......	10	...	10	80	100
Fine Yellow Brass.....	......	66	...	......	34	100
Gun Metal for Valves, etc........................	......	90	...	10	......	100
Journal Brasses for Rods, etc.................	......	80	...	17½	2½	100

In melting babbitt metal care must be taken not to burn it by overheating. Melt a part first in small chunks, and add remainder gradually. As soon as all melted, remove from fire and skim off the dirt. If heated beyond the melting point the softer components evaporate and leave the mass in a pasty condition.

When about to babbitt a journal wrap one thickness of common writing-paper smoothly around the bearing, fastening it in place with twine wound around in a regular spiral line three-sixteenths of an inch apart. The paper keeps the babbitt from getting chilled by the journal. It will, therefore, have a fine surface, and will also fit just right without any scraping. The twine leaves nice oil grooves.

Before pouring the metal through the oil-hole, make sure the journal is level and in central position. By means of two pasteboard rings fitting the journal, the ends of the box are closed, using putty or soft clay. A high funnel of clay is made around the oil-hole to facilitate the pouring in of the babbitt, and to increase the pressure, so as to have the babbitt fill the box perfectly.

www.ingramcontent.com/pod-product-compliance
Lightning Source LLC
Chambersburg PA
CBHW021157230426
43667CB00006B/444